Applied Condition Monitoring

Volume 19

The book series Applied Condition Monitoring publishes the latest research and developments in the field of condition monitoring, with a special focus on industrial applications. It covers both theoretical and experimental approaches, as well as a range of monitoring conditioning techniques and new trends and challenges in the field. Topics of interest include, but are not limited to: vibration measurement and analysis; infrared thermography; oil analysis and tribology; acoustic emissions and ultrasonics; and motor current analysis. Books published in the series deal with root cause analysis, failure and degradation scenarios, proactive and predictive techniques, and many other aspects related to condition monitoring. Applications concern different industrial sectors: automotive engineering, power engineering, civil engineering, geoengineering, bioengineering, etc. The series publishes monographs, edited books, and selected conference proceedings, as well as textbooks for advanced students.

** Indexing: Indexed by SCOPUS, WTI Frankfurt eG, SCImago

More information about this series at http://www.springer.com/series/13418

Fakher Chaari · Xavier Chiementin ·
Radoslaw Zimroz · Fabrice Bolaers ·
Mohamed Haddar
Editors

Smart Monitoring of Rotating Machinery for Industry 4.0

Springer

Editors
Fakher Chaari
Mechanics Modelling and Production Lab
National School of Engineers of Sfax
Sfax, Tunisia

Radoslaw Zimroz
Faculty of GeoEngineering Mining and Geology
Wrocław University of Technology
Wrocław, Poland

Mohamed Haddar
National School of Engineers of Sfax
Sfax, Tunisia

Xavier Chiementin
Institut de Thermique, Mécanique, et Matériaux (ITheMM EA7548)
University of Reims Champagne-Ardenne
Reims, France

Fabrice Bolaers
Institut de Thermique, Mécanique, et Matériaux (ITheMM EA7548)
University of Reims Champagne-Ardenne
Reims, France

ISSN 2363-698X ISSN 2363-6998 (electronic)
Applied Condition Monitoring
ISBN 978-3-030-79518-4 ISBN 978-3-030-79519-1 (eBook)
https://doi.org/10.1007/978-3-030-79519-1

This Springer imprint is published by the registered company Springer Nature Switzerland AG
The registered company address is: Gewerbestrasse 11, 6330 Cham, Switzerland

Contents

Vulnerabilities and Fruits of Smart Monitoring

Jablonski Adam and Tomasz Barszcz

Abstract "Smart" condition monitoring inherently implies that all other analysis techniques are "dumb". If so, how could one explain why for last half century, classical vibration-based condition monitoring techniques proved their merits in thousands of life-saving case studies? To discuss this concern, the paper briefly analyzes the process of evolution of condition monitoring systems over the years. For this purpose, the paper treats a condition monitoring system (CMS) as a part of a larger, much more complex system. The most important other systems CMS is connected to are the safety system, SCADA (Supervisory Control and Data Acquisition) and DCS (Distributed Control System). The outcome of such a complex system depends significantly on human actions (selection, configuration and operation), and the outcome of which serves other human actions (maintenance planning). Therefore, the paper tries to answer the question what is the actual "smartness" of modern systems that draws so much attention, namely is it the capabilities of smart systems or the hope in these capabilities? After reading this chapter, the reader would possibly gain some knowledge where to apply smart monitoring, and where do not.

Keywords Smart monitoring · Classical condition monitoring · Condition monitoring system

1 Introduction

1.1 The Ultimate System

In a perfect scenario, one would like to have a condition monitoring system, which just requires sensors mounting followed by pressing the "START" button or by plugging in the embedded system, and which provides completely reliable information about each machine part in a form like "Bearing degradation level: 77% (8 weeks to critical failure)". If so, why not connect this reliable system to the maintenance planning

J. Adam (✉) · T. Barszcz
AGH University of Science and Technology, Krakow, Poland
e-mail: ajab@agh.eu.pl

F. Chaari et al. (eds.), *Smart Monitoring of Rotating Machinery for Industry 4.0*,
Applied Condition Monitoring 19, https://doi.org/10.1007/978-3-030-79519-1_1

system, and to order parts and schedule repairs automatically? As well as it sounds, today it would be difficult to find any CEO that would agree to have a system which takes over financial strategy (turning it into potentially deadly scheme). It seems like on one hand industry more and more calls for "intelligent", "smart", "autonomous" systems, capable of automatized data collection and analysis, but on the other hand, manufacturers try to achieve this "smart" status with minimum modifications of currently offered systems, because these systems are reliable, effective, and most importantly verified. This paper therefore attempts to explain the actual meaning of "smart" system, how this "smartness" is achieved, and finally what consequences on the overall CMS performance "smartness" has. The paper has a conceptual character.

Smart CMS, by definition, aims in automation of all actions within condition monitoring, from which the machine-operator graphical interface draws most attention, simply because it is most eye-appealing, like demonstrated in Fig. 1.

The remaining parts, including selection and configuration stay in the shadow for the reasons given in the paper. Imagine a beginner that uses equipment, which gives information like "Large imbalance detected. Stop and fix." Probably, he would be very satisfied, and would order immediate repair. On the other hand, if it happened to an experienced machine operator, he would ask the system for velocity order spectrum. As a consequence, system's advanced diagnostic options (like data selection and spectrum display) are sometimes desired and sometimes detrimental. For many years, this observation led CMS manufacturers to prepare a large CMS portfolio,

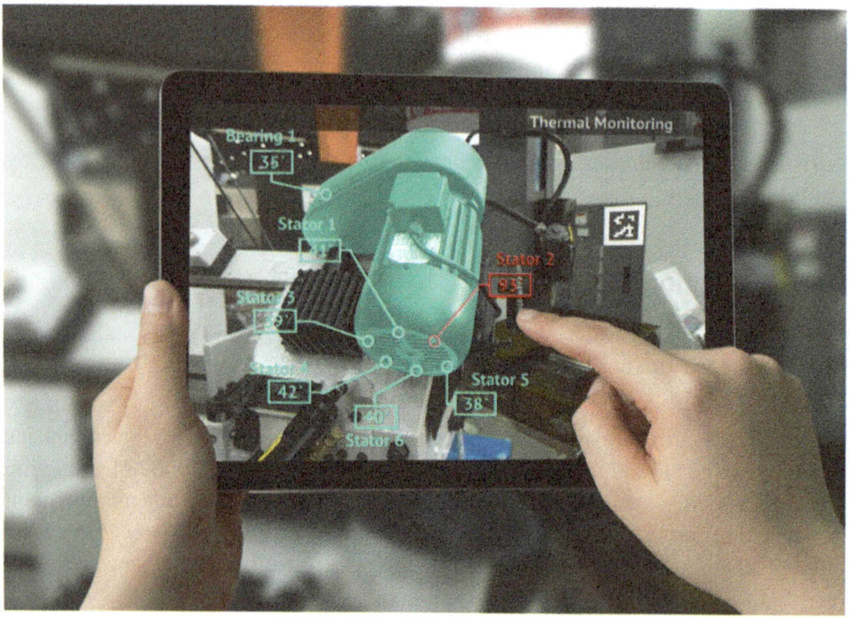

Fig. 1 Exemplary visualization of a smart CMS [available @ Allied Reliability_eBook_Industrial Evolution.pdf]

typically covering from basic, 1 or 2-channel devices with basic scalar diagnostic estimators, through portable data analyzers and wireless systems, to multi-channel distributed systems with separated modulus for data collection and data analysis. Naturally, over the years, many companies prepared platforms, which enable integration of data from any of listed types of equipment, like Emerson® Plantweb OpticsTM [1]. Other providers, like Allied Reliability® recommend external PTC ThingWorx platform [2].

1.2 What Is Smart Monitoring?

It is hard to tell, because nearly all currently available commercial condition monitoring systems claim that they are smart. For instance, Smart Condition Monitoring from Mitsubishi ElectricTM claims to create a "memory map" of a normal operating condition and to use "sophisticated algorithms" to detect abnormal state and offers "better understanding" of machine defect due to "higher level network". Simultaneously, GE™ states to use the same algorithms as "big data companies" analyzing the current behavior and past behavior of the plant. Allied Reliability™ promotes SMARTCMB™ as a system that is IIOT (Industrial Internet of Things) "ready", and that it increases uptime and decreases maintenance cost. Others, [3] emphasize the role of smartphones in enhancement of the effectiveness of condition monitoring systems for reliable machinery protection. Finally, some latest solutions like [4] refer to smart "on-site machine diagnostics" as an alternative to "***traditional*** cloud-based technology". Obviously, such contradictory scope might be a bit confusing.

1.3 Smart Systems Versus Smart Staff

Smart CMS offer "easier" installation and "easier" data analysis. In case of system commissioning, easier installation typically means more default settings within system configuration. Easier data analysis could be realized in two general ways. In the first case, automatized machine diagnostics is realized as a simple transformation of predefined data containers into descriptive information. For instance, the amplitude of shaft order could be tracked and converted to "Imbalance" level. The second general set of methods refers to Data Science analysis, like pattern recognition or ANN algorithms. In this case, the operator is somewhat compelled to "believe" in the system outcome. In both cases, smart systems inevitably subtly yet craftily remove skilled workers form individual partial actions within entire condition monitoring process. The key point is to analyze which steps of a human work could be efficiently replaced by a program, and which could not. Of course, the answer to this question is not simple; nevertheless, the answer that all the actions could be successfully replaced seems incorrect today. For practical goals, the paper shows few examples of successful implementation of smart methods in CMS. Worth mentioning, many diagnostic engineers from large companies complain that they regularly undergo shifting from one department to another, resulting in inability of mastering in a specific branch of technical science. Consequently, many machine diagnostic engineers do not have a solid background in classical condition monitoring methods; therefore, they tend to overestimate the capabilities of "smart" systems, believing them to be a perfect remedy to all their concerns.

2 Evolution of Condition Monitoring Systems

2.1 Early Days

First condition monitoring systems were developed for protection of high value assets, typically in power generation or chemical industries. The value of machinery and enormous costs of lost production (not mentioning the need to rebuild the plant itself) were so immense that it justified very high costs of development. As a result, the first condition monitoring systems were very expensive as well. The very first systems used analogue electronics, which was fast replaced by digital circuits.

Since the early days there was a distinction between monitoring and diagnostics. Monitoring (also referred to as protection) is a must for industrial machinery and is the first their functionality. Reaction of the system must be taken very fast and in a fully automated mode. It is necessary to react in milliseconds to an unexpected sudden event, for instance a broken turbine blade. In such a case the protected machine must be brought to stop before consecutive damage will happen. The second level is diagnostics, focused on early detection of faults. While protection systems only calculate few signal features, the diagnostics level involve calculating numerous advanced signal features, e.g. narrowband rolling bearing features. The system tracks trends of features and is able to detect early signs of technical state deterioration, even when the machine is still perfectly functional.

2.2 Expansion of Stationary Distributed Systems

Two major trends shaped the development of CMS, namely rapid development of digital technologies and—at the same time—equally rapid decrease of IT technology prices. Since many signal analysis methods were developed, standards (primarily ISO10816 and ISO7919) were needed to keep compatibility necessary to compare vibration levels between machines and systems. The protection systems began to proliferate into more and more assets. The distinction between the two layers became standard for critical machinery, e.g. power generation and oil and gas. It was adopted by standards (API670) which explicitly requires that these two layers should be separated into different computer systems. Moreover, failure of the diagnostics layer must not compromise the operation of the protection layer. Such a safety was achieved at the cost of more expensive CMS. In numerous other, less critical applications, where potential losses are smaller and fault development slower, the approach towards CMS reliability is not as demanding. It is common to mix the two functions in a single CMS. Dozens of manufacturers started to develop and offer much simpler (and less expensive) systems. These were installed in many other industries, starting from auxiliary machinery in critical plants, to transportation, food, marine to name only a few.

2.3 Industrial Internet-of-Things

The next big change was driven by further explosion of IT capabilities at continuously lower costs mixed with the advent of enhanced communication (including wireless). More and more machinery could be equipped with a CMS. Decreasing prices could justify smaller and smaller benefits (though still substantial). Other trends included cloud based systems, where the data from hundreds of CMS were sent, stored and analyzed by remote servers. The default tool to access the data became a web browser. Other consequence was also decreasing level of skills, as the vibration-based features were presented to normal machine operators, without any exposure to vibration analysis.

3 CMS Interaction with Human

3.1 Selection

The true meaning of a suitable selection of CMS is typically underestimated due to few reasons. Firstly, not many people are familiar with various types of such systems. If one works with portable equipment exclusively, he will seek for better portable equipment disregarding stationary systems, and vice-versa. Secondly, CMS are selected by management staff on the basis of business plans, which generally boils down to cheapest systems. In this case, the idea is that any CMS is equally good for the job. Thirdly, in many plants, the equipment is partially or totally inherited, which limits potential changes, because in nearly all cases, systems from different manufacturers are not compatible. As a result, in many applications, systems are not suitable for any significant improvements permanently from the start. As a very common example of unsuitable selection of CMS elements one could consider a set of acceleration sensors with 100 mV/g sensitivity for a high-volume machine with a relatively large transmission ratio, for which the vibration level between front and back end easily differs by more than order of magnitude. Situation where suitable sensors with higher sensitivity are installed at locations with smaller vibrations are rarely met in practice. Therefore, it might be concluded that fundamental rules of selection of suitable CMS for individual scenario should be followed prior to consideration of system "smart" features. Other words, it is NOT recommended to select a system which suits ones needs from available smart systems, but rather to look for a suitable system without adding any initial value to "smart" class of system.

3.2 Configuration

Among various actions, which refer to the process of machine condition monitoring, configuration of the system is a major taboo—it is skipped, it is depreciated, and it is disliked. This popular approach, which underestimates the meaning of CMS configuration, is like a minefield, because configuration decides what data is processed, when it is processed, and how it is processed. Moreover, configuration process itself is long and costly, yet it does not bring any direct benefit to the operator (or the plant), so it is treated as a necessary evil. As a result, configuration is omitted during business system presentation, and is shifted to support actions. During the first training, frequently it is found to be much more troublesome than system operation. For beginning users, the less optional the configuration the better. For more advanced users, it is just the opposite. As a result, it is very difficult to provide a configuration interface, which would satisfy a large number of users.

Configuration is typically divided into few phases. First phase refers to system pre-configuration, which is done by the manufacturer and it is exaggerated to make place for further adjustment. For stationary systems and advanced portable systems, initial configuration also includes definition of drive train kinetostatics (frequently called "kinematics") and narrowband analyses. Each narrowband analysis includes a configuration subset, which covers spectrum type, spectral range, optional filters, amplitude type (peak, root-mean square, power, sum), etc. In the third phase, data is additionally classified into operational states, so that vibrations only in similar machine dynamic states are compared.

Definitely, successful replacement of human actions within CMS configuration process is exceptionally attractive. But what exactly would it mean when each configuration element is selected individually? Selection of sampling frequency is generally fixed, so is the length of signals. The location of each sensor is taken either from norms or from human experience. Next, almost all commercially available systems automatically calculate narrowband analyses on the basis of MANUALLY prepared kinetostatic configuration. Is it possible to further automatize any of these parts? So far, it is noticed that "smart" configuration features are limited to simple actions, like automatic determination of shaft-related analyses on the basis of the phase marker (PM) signal or automatic triggering for data storage. An interesting solution for automatic threshold configuration for scalar trend analyses could be found in a modern AVM4000 system [5], which is based on percentile limits of cumulative distribution functions. It might be therefore concluded that smart systems should prepare large configurations automatically, but this approach might not be correct at all. Alternatively, large static configuration of a system could be skipped, as long as the system is not expected to give fault identification, i.e. just fault detection and possibly fault severity assessment.

3.3 Operation

Operation of CMS refers to the direct interaction of a machine operator with the system and it is composed of different elements depending on the system architecture. For unsupervised protection systems, desired system interaction is none. For simple portable systems, data acquisition is triggered, followed by internal signal processing. The displayed data is analyzed by the operator on-site. In case of stationary distributed systems, data is transferred to some central unit, to which a diagnostic engineer is connected. Desirably, such systems operate on events, which are signals to the engineers that machine needs attention on the basis of the current data. From the operation point-of-view, smart system could refer to two aspects, namely data transfer and data analysis. Firstly, in any of mentioned systems, a smart system could be connected to some network enabling automatic data transfer. This is especially attractive to portable systems, where such feature significantly saves time.

Secondly, in case of portable and stationary monitoring systems, smart operation refers to automatic data analysis. This data analysis answers three fundamental questions:

1. Is there (a new) machine fault?
2. What is the fault element?
3. How serious is the fault?

The first question refers to fault detection, the second to fault identification, while the third one to severity assessment. In case of a smart vibration-based condition monitoring system, the first concern is realized by unsupervised anomaly detection. In this scenario, a classically "permissible" machine technical state is classified as a "normal" state, while any significant deviation from this state is called an "anomaly" or "abnormal" state. Although a commonly accepted classification of vibration-based data science methods does not exist so far, in this paper it is accepted that "machine learning" covers all unsupervised methods, which operate on predefined scalar diagnostic estimators (also called "health indicators" HI or signal "features"), while "deep learning" refers to all unsupervised methods operating on raw vibration data.

Unsupervised analysis based on scalar diagnostic indicators is a bit tricky. Before any of such analysis is done, it needs to be stated that three types of indicators exist. The first group is wideband indicators, which means that they cover "entire" signal in some domain. These indicators include peak-to-peak (PP), root mean square (RMS), crest factor, and kurtosis, from both, acceleration and velocity signals. The second are narrowband indicators, typically calculated in frequency (or order) domain. The third set refers to indicators, permissible values of which are to be found in norms (like velocity RMS from ISO 20816). Starting from the last group, the verification of permissible vibrations is straightforward; therefore, smart analysis seems to be pointless. In case of narrowband indicators, the set is limited, and so is the anomaly detection capability. For wideband indicators, the number of analysis is relatively small, so it is easy to handle them in a classical way.

Recalling the configuration process described in previous section, note that in a classical CMS, for every diagnostic indicator, the system stores permissible Warning and Alarm levels, which generate an event upon trespass. These considerations generate following deduction: if one is able to define diagnostic indicators and corresponding threshold levels correctly (classical way), the system should react properly on the change of the technical condition of the machine; if one is not able to do so, then why believe that more advanced, smart, unsupervised machine learning methods would work at all?

3.4 Maintenance Planning

Every CMS has the very same ultimate purposes, i.e. to protect life and to reduce production costs by providing information about (degradation of) technical condition of the machine. For machine protection systems, this information is sent directly to a SCADA system, and it has a form of a control electrical signal. For the rest of vibration-based systems, this information could be described by its form (high-resolution graph, embedded bar graph, display value, sms, e-mail, sound, light, etc.), its reliability (formalized as "false alert rate"), and its content (numerical value, shape of the graph, text description, pictogram, color change, etc.). For classical CMS, these parameters are well established and well understood, and it might be hence concluded that any improper performance of such system is caused by improper (faulty or incomplete) system configuration or data corruption. For instance, overestimated threshold levels would fail to detect fault. For smart systems, each of described parameters is somehow difficult. Results of many smart methods are in a form of some numerical "rate", which have connotations with the data, but not with machine elements. The reliability of such methods is hard to determine, because typically they do not operate on predefined scalar threshold level, which requires a subsequent interpreter, which generates clear information. Without such interpreter, it could easily happen that simple set of information generated by a classical architecture would be transformed by a smart system into elaborated, equivocal data.

4 Recommendations for Selection of Suitable System

If the reader has arrived that far in the chapter, the natural reaction would be to ask, WHAT is thus the optimal CMS? It is a proper question, but the answer is quite complex. The selection process is a result of two prior questions, namely what is the monitored machine and what are its failure modes? The first one is whether we need a protection layer or only diagnostics? Are the simplest signal features like rms sufficient or do we need a complex set of dozens of features? The second question should answer what is the level of expertise of the system users? The "smartness" of CMS should first focus on efficiency of commissioning, i.e. installation and configuration. Then, the system should provide timely and sufficient information to its users. As the popular saying goes, it should be as simple as possible, but not simpler.

5 Summary

The paper starts with a concept of a perfect "smart" vibration-based condition monitoring system. Up to now (to the authors' knowledge), a system which fulfills all the customer needs does not exist. Moreover, there is not any known theory that would justify that it is possible to design a fully automatized CMS. Yet, as given in the paper, CMS providers are racing towards "game changing" systems claiming systems' smartness where possible. At the same time, it could be found in [6] that regardless of the CMS type, only 5% of collected data is actually analyzed in industrial environment, because the rest of the data is insignificant or corrupted. More details of corrupted data handling are found in [7]. Therefore, the final conclusion from the paper is that although "smart" condition monitoring offers many attractive fruits, it is much more vulnerable to inexperienced, new equipment specialists than classical systems.

Acknowledgements The paper is financially supported by The National Centre for Research and Development, grant No. POIR.04.01.04-00-0080/19 (BLASTER).

References

1. https://www.emerson.com/en-us/automation/asset-management/asset-monitoring/health-monitoring/plantweboptics
2. https://www.ptc.com/en/resources/iiot/product-brief/thingworx-platform
3. Mark S (2019) Smart monitoring for the intelligent machine age. Bently Nevada, ORBIT ARTICLE, Baker Hughes. (https://www.bakerhughesds.com/orbit-article/smart-monitoring-intelligent-machine-age)
4. https://www.youtube.com/watch?v=CDUZaW8Cxa8
5. http://amcvibro.com/product/avm-4000-3/
6. https://www.industr.com/en/online-monitoring-a-deeper-insight-into-asset-health-2474814
7. https://news.usni.org/2019/09/19/usni-news-video-navy-asks-civilians-to-solve-persistent-problems-in-weekend-hack-a-thon

A Tutorial on Canonical Variate Analysis for Diagnosis and Prognosis

Xiaochuan Li, Tianran Lin, and David Mba

Abstract Canonical variate analysis is a family of multivariate statistical process monitoring tool for the analysis of paired sets of variables. Canonical variate analysis has been employed to extract relations between two sets of variables when the relations have been considered to be non-linear, when the process is non-stationary and when the dimensionality needs to be reduced to benefit human interpretation. This tutorial provides the theoretical background of canonical variate analysis. Together with the industrial examples, this study discusses the applicability of the extensions of canonical variate analysis to diagnosis and prognosis. We hope that this overview can serve as a hands-on tool for applying canonical variate analysis in condition monitoring of industrial processes.

Keywords Canonical variate analysis · Diagnosis · Prognosis

1 Introduction

When a process can be described by two sets of data corresponding to two different views, investigating the relations between these two aspects may provide new information about the functioning of the system. The relations refer to as a mapping of the variables of one aspect to the variables of the other aspect. For instance, in the field of medicine, one aspect could be related to the symptoms of the disease and the other corresponds to the risk factors that could affect the disease. Investigating the relations between the symptoms and the risk factors can provide more information on the disease exposure so as to give advices on treatment and cure. These relations can be studied by means of canonical variate analysis that has been developed for this purpose.

X. Li (✉) · D. Mba
Faculty of Technology, De Montfort University, Leicester L1 9BH, UK
e-mail: Xiaochuan.li@dmu.ac.uk

T. Lin
School of Mechanical and Automotive Engineering, Qingdao University of Technology, Qingdao, China

F. Chaari et al. (eds.), *Smart Monitoring of Rotating Machinery for Industry 4.0*,
Applied Condition Monitoring 19, https://doi.org/10.1007/978-3-030-79519-1_2

Canonical variate analysis (CVA) is a family of multivariate statistical process monitoring (MSPM) tool. CVA's MSPM counterparts include principal component analysis (PCA) [7], independent component analysis (ICA) [10] and partial least-squares analysis (PLSA) (Kruger and Dimitriadis 2008), etc. These basic MSPM methods perform well under the assumption that process variables are time independent. However, this assumption might not hold true for real industrial processes since sensory signals affected by noise and disturbances often show strong correlations between the past and future sampling points [11]. Therefore, variants of the standard MSPM approaches [20, 29, 34] were developed to solve the time-independency problem, which makes these approaches more suitable for dynamic process monitoring. In addition to approaches derived from PCA, ICA and PLSA, canonical variable analysis (CVA) is a method that can explore the relations between the systems' past and future status, thereby making it a strong candidate for dynamic process monitoring.

Although CVA is a linear model, which means it may lead to problems in monitoring systems that generally operate under time-varying conditions, but it has also proven to be high performing in this context if properly managed or modified [14, 25, 28]. This assumption makes it interesting for applications in contexts such as chemical process that operate with different characteristics in the various processing cycles, and large-scale rotating machinery such as compressors and gas turbines that involve switch between high and low working loads. The multivariate statistics technique of canonical variate analysis allows, unlike other multivariate statistical techniques, to consider the time dependence of variables during process monitoring. In fact, CVA can properly identify features and dynamic information of the time series allowing to find the maximum correlation between past and future measurements [33]. This characteristic makes it a very suitable technique for real monitoring applications because the assumption of independence from the time of variables is often wrong in real production processes [11]. Moreover, dimensionality reduction techniques allow increasing the ability to identify a fault, also increasing the adaptability to new data of what is proposed [26]. The important role of CVA is even based on the consideration that industrial processes consist of a large number of process variables operating at controlled conditions and it could be useful to consider a state-space realization in such processes [21]. It has also been confirmed that dynamic models constructed by CVA demonstrate higher accuracy compared to dynamic PCA in terms of diagnostics [21]. Time-varying characteristics certainly make condition monitoring more complex. This means that the state of health of the process, as well as the presence of a fault, is manifested with values of the parameters of processes that are not constant.

Existing CVA-based methods that have been developed to address the problem of fault diagnosis can generally be divided into four categories: (1) traditional CVA and its linear variants. Their applications in industrial processes can be found in [13, 22]. Apart from CVA, its variant—Canonical Variate Dissimilarity Analysis (CVDA)—which was designed to improve CVA's ability to detect incipient faults, has been proven to be effective and superior to the traditional CVA method as stated in [24]. (2) kernel CVA, which was developed to further improve the diagnosis performance of CVA in the presence of system non-linearities [23], (3) Adaptive CVA. Adaptive

CVA was developed for the monitoring of dynamic processes where variations in operating conditions are incurred [14]. (4) Just-in-time-learning based CVA (JITL-CVA). JITL-CVA was proposed due to the reason that CVA has deficiencies in handling processes with multiple operating points. JITL-CVA has been proven to have better fault detection performance that its CVA counterparts while still tracking changes in the system [6]. This study will discuss in detail how CVA is utilized for fault diagnosis, and its extension CVDA method will also be reviewed. This tutorial also discusses how the key tunning parameters are determined through numerical examples.

Unplanned downtime caused by system failure is costly and can incur large economic losses and security threats. As a result, predictive maintenance has been a very active field of research in recent years, where system failures are estimated, and maintenance is implemented on an as-needed basis. However, it is difficult and costly to carry out remaining useful life (RUL) prediction when equipment is under normal conditions since little information about the degradation trend can be found during this stage. To make a prognostic framework suitable for online monitoring, it is essential to include a module which can automatically determine prediction start time such that the RUL prediction is implemented only after certain failures are detected. CVA can act as a good starting point for prognosis since CVA based monitoring index which is based on the deviations between past and future measurements can be adopted to automatically determine the prediction start time. Additionally, the constructed monitoring index provides valuable information about the health status of the equipment, and therefore can be used to predict the RUL. Furthermore, CVA can be considered as part of the context of the data-driven models, where an a priori knowledge of the physical structure of the considered context is not necessary. In this type of model, the available data and their history is manipulated and studied in order to transform it into useful knowledge for the fault diagnosis process, and consequently for the subsequent decision-making process [1]. This process is called "feature extraction" and can be performed with different approaches, the main ones being the qualitative approach, such as expert qualitative, and quantitative, such as PCA [30–32]. The contribution presented in this paper for the decision-making process is strictly representative of a quantitative approach. Numerous are the contexts in which the decision-making process and its correlation with the fault detection process has been analysed. The theme highlighted in each of these contributions is the need for the decision-making process to be effectively cost-effective, ensuring that condition monitoring do not involve unnecessary maintenance operations, with consequent unnecessary costs [2]. The same contribution highlights how the joint analysis of current and past condition monitoring allows, in combination with other elements, to improve the maintenance decision-making process. This is closely related to the technique considered in this paper, i.e. CVA, because as previously mentioned it allows to evaluate the temporal relations between input data.

Generally speaking, CVA-based prognostic method can be categorized into two different groups: (1) CVA state space model. CVA itself is a state-space model-based method, and its output can be used to build a state-space model that represents the dynamics of the system [15]. (2) CVA-based data driven models. The output of CVA can serve as a condition indicator of the system, and this condition indicator is often utilized in combination with predictive data driven models to form a prognostic scheme [17, 19].

This tutorial starts with an introduction to the original formulation of CVA. The basic framework and techniques for determining optimal parameters are discussed. The variants of CVA, including CVDA and adaptive CVA, are illustrated using industrial case studies. The tutorial also discusses the state space-based CVA and CVA-data driven methods for systems prognostic analysis. This tutorial acquaints the reader with canonical variate methods, discusses where they are applicable and what kind of information can be extracted.

2 Canonical Variate Analysis for Diagnosis

2.1 The Basic Framework of CVA

The aim of this section is to discuss the mathematical procedure for the application of CVA. The fundamental steps for the application of CVA, following what presented in [13], are presented in this section.

In the case of CVA, the variables of an observation can be partitioned into two data sets that can be seen as the two aspects/views of the data. CVA application for fault detection has been proposed in 2010 [22], where the two datasets to correlate are the past and the future data that are created from the measurements in maintenance application. In this tutorial, we assume that the observations are standardized to zero mean and unit variance. The main aim of CVA is to extract and find the maximum linear relations between the two views.

We consider the multivariate measured data $y_t \in R^n$, R^n, where the process variables are n and t is the sampling time. In order to generate two data matrices from this measurement, we expand each sampling including p past samples and f future samples, with the rule $p = f$. Then the future and the past samples vectors $y_{f,t} \in R^{fn}$(e.g. $y_{f,t}$ is a vector of size fn) and $y_{p,t} \in R^{pn}$

are obtained as follows

$$y_{f,t} = \left[y_{t+1}, y_{t+2}, \ldots, y_{t+f-1}\right]^{\mathrm{T}} \tag{1}$$

$$y_{p,t} = \left[y_{t-1}, y_{t-2}, \ldots, y_{t-p}\right]^{\mathrm{T}} \tag{2}$$

$y_{f,t}$ and $y_{p,t}$ are then normalized to mean zero vectors $\hat{y}_{f,t}$ and $\hat{y}_{p,t}$ to avoid the dominance of process variables with large absolute values.

The first step generates two data matrices from the measured data $y_t \in R^n$, where the process variables are n and t is the sampling time. This normalization manages and avoids an excessive influence on the monitoring result of variables with larger absolute values. $\hat{y}_{f,t}$ and $\hat{y}_{p,t}$ are arranged in columns for creating the future observations matrix $\hat{Y}_f \in R^{fnXN}$ and the past observations matrix $\hat{Y}_P \in R^{pnXN}$ where $N = m - p - f + 1$ and m represents the number of total samples in y_t and t assumes past and future values.

$$\hat{Y}_f = \left[\hat{y}_{f,t+1}, \hat{y}_{f,t+2}, \ldots \hat{y}_{f,t+N}\right] \tag{3}$$

$$\hat{Y}_p = \left[\hat{y}_{p,t+1}, \hat{y}_{p,t+2}, \ldots, \hat{y}_{p,t+N}\right] \tag{4}$$

Constructing the past and future matrices in this way allows each column to have information from the nearest f/p samples and each row to have measurements in a chronological order. $\hat{Y}_f$ and $\hat{Y}_p$ are the two aspects from which we would like to evaluate the original observations. The principle behind CVA is to find two positions in the two data spaces respectively that have images on a new coordinate system such that the correlations between them is maximized. The positions of the two data sets can be obtained through techniques of functional analysis. Commonly used functional analysis include the eigenvalue-based methods [9], solving the generalised eigenvalue problem [4] and the singular value decomposition (SVD) [8]. In this tutorial, we discuss solving CVA through singular value decomposition.

The SVD method starts with computing the variance matrices of $\hat{Y}_f$ and $\hat{Y}_p$. According to Samuel and Cao [27], due to the symmetric positive definite property, the square root factors of the matrices can be found using a Cholesky or eigenvalue decomposition. By applying the Cholesky decomposition to $\hat{Y}_f$ and $\hat{Y}_p$, a Hankel matrix H can be formulated as per (5). The SVD decomposes the Hankel matrix H to find the linear combinations that maximizes the correlation between $\hat{Y}_f$ and $\hat{Y}_p$.

$$H = \Sigma_{ff}^{-1/2} \Sigma_{f,p} \Sigma_{p,p}^{-1/2} = U \Sigma V^T \tag{5}$$

In (5):

- $\Sigma_{f,f}$ and $\Sigma_{\mu,\infty}$ are the covariance matrices of $\hat{Y}_f$ and $\hat{Y}_p$
- $\Sigma_{f,p}$ is the cross-covariance matrix of $\hat{Y}_f$ and $\hat{Y}_p$

Defining r as the order of H, one can define U (6), V (7), and Σ (8) as:

$$U = [u_1 u_2, \ldots, u_r] \in R^{npXr} \tag{6}$$

$$V = [v_1, v_2, \ldots, v_r] \in R^{nfXr} \tag{7}$$

$$\Sigma = \begin{bmatrix} d_1 & \cdots & 0 \\ & \ddots & \\ 0 & & d_r \end{bmatrix} \in R^{rXr} \tag{8}$$

The columns of U are called left-singular vectors of H and the columns of V are called right-singular vectors of H. n is the number of process variables. Σ is a diagonal matrix, in which the diagonal elements are called singular values, representing the degree of correlation between the values in U and V. The positions of past and future data $\hat{Y}_f$ and $\hat{Y}_p$ are obtained from $V^T \Sigma_{p,p}^{-\frac{1}{2}} * \hat{Y}_p$ and $V^T \Sigma_{f,f}^{-\frac{1}{2}} * \hat{Y}_f$. But when CVA is applied for fault detection, a common practice is to further partition the positions into a state space and a residual space, respectively. In order to do so, Σ is subsequently used to truncate the matrix V to V_q (9), using the largest q singular values permit to truncate, and the details of this process will be illustrated in Sect. 2.2.

$$V_a = \left[v_1, v_{2,}, \ldots, v_a\right] \in R^{npxq} \tag{9}$$

$\hat{Y}_p$ is then converted in a reduced q-dimensional matrix $\zeta \in R^{qXn}$ (10) based on V_a, ζ is also referred to as the image of the position $V^T \Sigma_{p,p}^{-\frac{1}{2}} * \hat{Y}_p$.

$$\zeta = \left[Z_{t=1} Z_{t=2,}, \ldots, Z_{t=N}\right] = K^* \hat{Y}_p \tag{10}$$

$$K = V_q^T \Sigma_{p,p}^{-1/2} \in R^{qXnp} \tag{11}$$

The residual space ψ is computed as (12).

$$\psi = [\varepsilon_{t=1} \varepsilon_{t=2}, \ldots, \varepsilon_{t=N}] = G * \hat{Y}_p \tag{12}$$

$$G = I - V_q V_q^T \Sigma_{p,p}^{-1/2} \in R^{npXnp} \tag{13}$$

Similarly, ψ is the image of the position $V^T \Sigma_{f,f}^{-\frac{1}{2}} * \hat{Y}_f$. In (10) and (12) * means multiplication.

ζ and ψ contain the vectors z_t and ε_t being placed in a descending order, which represent the canonical correlations between the past and future matrices in a descending order. This is because vectors z_t and ε_t correspond to the elements in matrix V, which captures the canonical correlations between the past and future data through SVD. It is obvious that ζ and ψ are obtained through applying the SVD and the calculations illustrated in (10)–(13). ζ represents the largest q canonical correlations, while ψ include the system dynamics are not captured by ζ.

ζ and ψ together fully capture the system dynamics and can therefore be utilized to construct a health indicator that indicates the system status for monitoring purposes.

The health indicators are constructed with the canonical variates z_t and residual variates ε_t. Commonly used health indicators are the Hotelling T^2 (14) and Q (SPE) (15) statistics [9]. The Hotelling T^2 and Q (SPE) are calculated as follows

$$T_t^2 = \sum_{j=1}^{q} z_{t,j}^2 \tag{14}$$

$$Q_t = \sum_{j=1}^{np} \varepsilon_{t,j}^2 \tag{15}$$

T^2 takes into account the projection of the measurement matrix into the q-dimensional space, and Q captures the system variations not considered by T^2. In terms of fault diagnosis, it is possible to assert that the system is in a faulty condition when the health indicator surpasses a normal threshold, computed with the data representing the normal operative behavior of the system.

2.2 *Determination of the Number of Retained States*

An important parameter that affects the diagnosis results is q, i.e. the value that determines the proven truncation of the matrix V. Different methodologies have been proposed for the computation of the retained states q. The techniques most popular are those based on considering the dominant singular values in the matrix Σ. The "knee point" method was put forward recently, and the basic assumption of this method is that one can estimate q to be the point where a "knee" appears in the singular values curve [24]. Figure 1 illustrates how a knee point can be found in the singular value plot. Another method that is based on dominant singular values is cross-validation [18]. The principle of this method is to find the optimal q through

Fig. 1 "Knee point" method

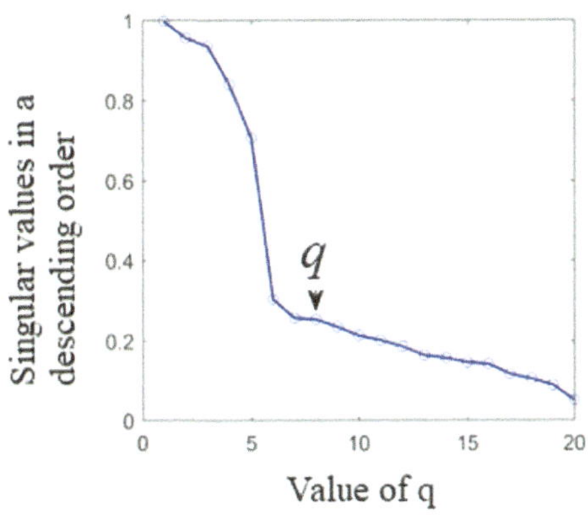

minimizing false alarms during cross-validation. This method can largely guarantee the false alarm rate when new measurement becomes available. Akaike Information Criterion has also been adopted to determine the value of q [3].

2.3 Determination of Fault Threshold

To recapitulate, CVA can be applied to transform process data into a one-dimensional health indicator and consequently can be used to monitor machinery performance and performance fault diagnosis. This monitoring approach has two stages:

- A first training phase, in which data related to the machinery during normal healthy operating conditions are used to define a threshold of normality of the indicator.
- Then there is the actual process monitoring phase. A state of non-health of the machinery is identified when the machinery health indicator exceeds the previously calculated threshold.

Since real-world measurements are non-Gaussian processes, fault thresholds for real-time monitoring need to be computed through a non-Gaussian approach. One commonly used solution is to estimate the probability density function directly for these health indicators through a nonparametric approach called kernel density estimation (KDE) [5]. The KDE is a well-established method to estimate the probability density function for univariate variables, thereby making it particularly suitable for the estimation of the threshold of CVA health indicators.

Given the probability density function $p(x)$ of a random variable x, the probability that x is smaller than a specific value c is calculated as follows:

$$P(x < c) = \int_{-\infty}^{c} (x)dx \tag{16}$$

The estimation of the PDF, $\hat{p}(x)$, of x through kernel Gaussian estimation is given by the following:

$$\hat{p}(x) = \frac{1}{N \cdot BW} \sum_{k=1}^{N} K\left(\frac{x - x_k}{BW}\right) \tag{17}$$

where N refers to the number of samples of variable x. BW is the selected bandwidth of KDE. There is no single perfect way to calculate the BW. However, as suggested in [22], a rough estimation of the optimal BW can be described as follows:

$$BW = 1.06\sigma N^{-0.2} \tag{18}$$

where σ is the standard deviation, and N is the number of training data points being taken into consideration. The kernel function utilized in this study is given by the following:

$$K(w) = \frac{1}{\sqrt{2\pi}} e^{\frac{w^2}{2}} \tag{19}$$

By replacing random variable x with Hotelling's T^2 and Q statistics, the thresholds for both health indicators are calculated from the PDFs of $T^2$and Q health indicators for a given critical level, α, by solving the following formulas:

$$\int_{-\infty}^{T_\alpha^2} P(T^2) dT^2 = \alpha \tag{20}$$

$$\int_{-\infty}^{Q_\alpha} P(Q) dQ = \alpha \tag{21}$$

T_α^2 and Q_α are the thresholds for the Hotelling's T^2 and Q statistics, and are the values that we would like to calculate. The CVA procedure for fault detection is summarized in Fig. 2. The main purpose of the offline training stage is to compute the thresholds. The online monitoring involves constructing the Hotelling's $T^2$and Q statistics for test data and comparing them with the thresholds calculated in the offline stage.

2.4 Extensions of CVA—Canonical Variate Dissimilarity Analysis

The traditional CVA T^2 and Q health indices may not be sensitive enough for incipient faults [24]. This is due to the reason that T^2 and Q statistics only assess the variations from one aspect of the original measurement, although the CVA itself maximizes the correlations between the past and future sets. A new health index, namely canonical variate dissimilarity analysis (CVDA), was proposed to assess the dissimilarity between the past and future canonical variates for indicating system health. The health index adopted by CVDA method is calculated as below.

Motivated by the fact that CVA is able to find the maximum correlations between past and future sets, one can detect small shifts by investigating how far away future canonical variates are deviated from past. The CVDA index that quantifies the distinctions between the past and future sets is computed as

$$r_t = G_q^T \hat{y}_{f,t} - \Sigma_q K_q^T \hat{y}_{p,t} \tag{22}$$

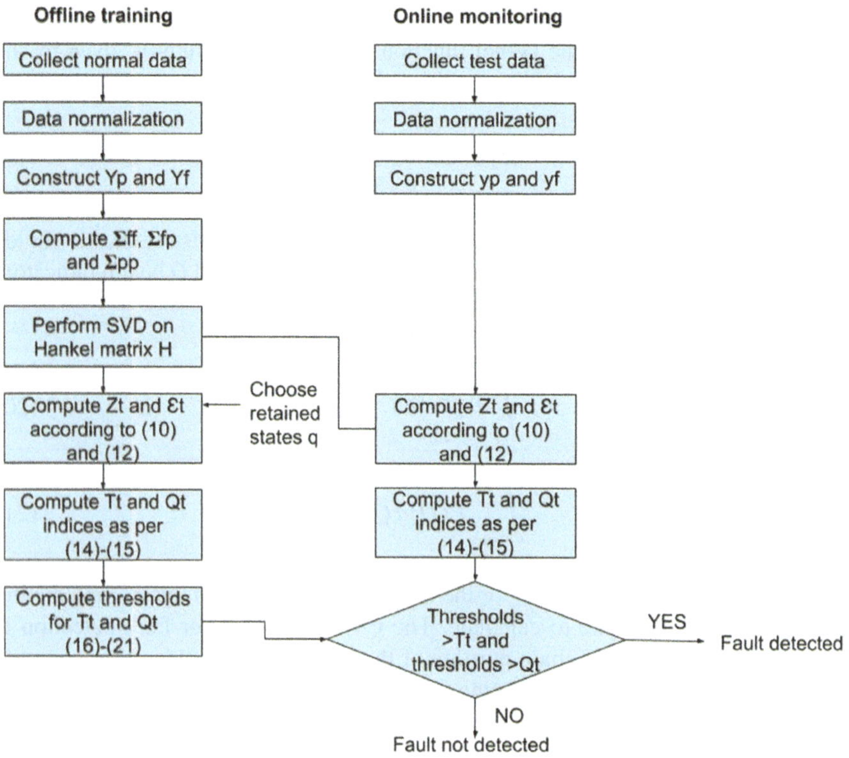

Fig. 2 CVA working principles

$\Sigma_q = \mathrm{diag}(\lambda_1, \lambda_2, \ldots, \lambda_q)$ is a diagonal matrix with its diagonal elements being the first q canonical correlations. r_t measures the discrepancies between the past and future sets and are able to better represents small changes in the system at early stage of faults, compared with health indices derived from traditional CVA approach [12].

The covariance of r_t can be calculated as:

$$\begin{aligned}\Sigma_r = E\left(rr^T\right) &= G_q^T E\left(\hat{y}_{p,t}\hat{y}_{p,t}^T\right)G + \Sigma K_q^T E\left(\hat{y}_{f,\cdot}, \hat{y}_{f,t}^T\right)K_q^T \Sigma^T \\ &\quad - G_q^T E\left(\hat{y}_{p,t}\hat{y}_{f,t}^T\right)G_q^T \Sigma^T - \Sigma K_q^T E\left(\hat{y}_{f,t}\hat{y}_{p,t}^T\right)G \\ &= I + \Sigma\Sigma^T - \Sigma\Sigma^T - \Sigma\Sigma^T = I - \Sigma\Sigma^T\end{aligned} \tag{23}$$

The distinctions between the past and future measurements are centred around a zero mean under healthy conditions. Hence, a diagnostic health index can be formed as the squared Mahalanobis distance of the discrepancy features from zero (i.e. by computing the sum of squares of the dissimilarities r_t for each time point t, standardized by the covariance matrix Σ:

$$T_d = \left[r_t^T\left(I - \Sigma\Sigma^T\right)r_t\right]^{1/2} \tag{24}$$

Apart from the health index, the CVDA and traditional CVA method share the same working process, including the construction of past and future sets (1)–(13), the determination of the number of retained states and the calculation of fault thresholds.

2.5 Industrial Case Study—Canonical Variate Analysis

The case study concerns a water treatment plants used to purify the water before entering the production cycle. The failure management of this system is essential because an incorrect operation of a WTP implies an injection of unpurified water into the production line. This improper release leads to production losses and significant economic damage caused by the potential disposal of the production batch. During a reverse osmosis process, the water is pushed onto the membrane by a pump, which exerts a higher pressure than the osmotic pressure. The pressure required to overcome the osmotic pressure depends on the concentration of the feed water: the greater this concentration, the greater the pressure required. A reverse osmosis system uses cross-filtration, more commonly called tangential filtration. In this context, contaminants are collected within the filter surface. To prevent the accumulation of contaminants, crossflow filtration allows the water to sweep away the accumulation of contaminants and induces turbulence strong enough to keep the membrane surface clean. The water purification guarantees that the characteristics of the products are maintained constant and do not compromise the subsequent operations that will be carried out on the product.

Following the traditional CVA method explained in Sects. 2.1–2.3, the plant condition monitoring data captured during healthy operating conditions lead to the training of the model and to the computation of the fault thresholds. After that, the data referred to the degradation process have been used to validate the trained model. The dataset used for the training contains 577 measurements and the faulty process contains 170 measurements. The tuning parameter q was set to 25 following the cross validation method [13].

Figures 3 and 4 provide process monitoring based on the T^2 and Q index respectively. As can be seen from both figures, both health indices successfully detect the fault at around 460th sampling point, given that the past p window length was set to

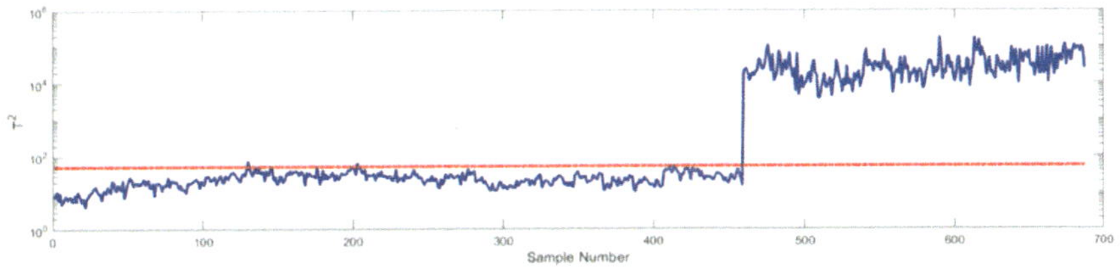

Fig. 3 Diagnosis results of T^2 index

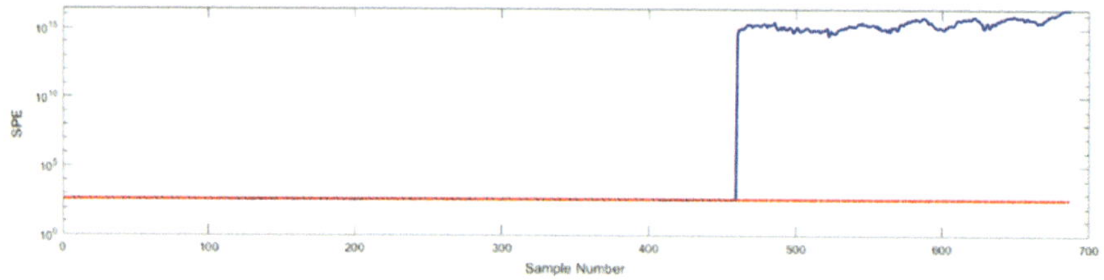

Fig. 4 Diagnosis results of Q index

59 (e.g. the actual detection happens at $59 + 460 = 519$th sampling point), making the total error rate for the fault detection process the same both for T^2 and Q, i.e. 7.8%, thus making it reasonable to assert that both indexes have the same fault sensitivity. All errors found are attributable to false alarms and not to underestimated dangerous situations.

3 Canonical Variate Analysis for Prognosis

CVA has been applied for both diagnosis and prognosis purposes. A few researchers have developed exploratory studies in prognosis by using the CVA method. We divide them into two categories, i.e., CVA-based data driven models and CVA state space-based models.

3.1 CVA-Based State Space Models

Apart from being a two-view multivariate statistical approach, CVA is also a state space model. CVA can be used to build a state space model that represents the dynamics of the system using condition monitoring measurements. This method requires that the system has one or more input (e.g. manipulated/controlled variables) variables that can be estimated ahead of time by looking at production plan or system control settings. The ideal behind this method is, given the future system input and past sensory measurements, to predict future system behavior through the CVA state space model.

We denote system input as u_t and sensory measurements as y_t, the CVA linear state space model can be built as follows [26]:

$$x_{t+1} = Bx_t + Cu_t + w_t \tag{25}$$

$$y_t = Dx_t + Eu_t + Lw_t + v_t \tag{26}$$

where x_t is the system states, B, C, D, E and L are model coefficient matrices; And w_t and v_t are independent white noise. According to the literature [22], canonical variates calculated through CVA (i.e. (10)) or CVDA (i.e. (22)) can be used to replace x_t if the number of retained states q is no less than the actual order of the system. The unknown coefficient matrices B, C, D and E can be computed through multivariate regression as follows

$$\begin{bmatrix} \hat{B} & \hat{C} \\ \hat{D} & \hat{E} \end{bmatrix} = \mathrm{Cov}\left[\begin{pmatrix} z_{t+1} \\ y_t \end{pmatrix}, \begin{pmatrix} z_t \\ u_t \end{pmatrix}\right] \cdot \mathrm{Cov}^{-1}\left[\begin{pmatrix} z_t \\ u_t \end{pmatrix}, \begin{pmatrix} z_t \\ u_t \end{pmatrix}\right] \tag{27}$$

where z_t represents canonical variates calculated through CVA (i.e. (10)), and in the case of using CVDA, z_t should be replaced with the system dissimilarity r_t (i.e. (22).

The procedures of performing system behavior prediction is described as follows.

- Obtain the system inputs u_t (usually contains manipulated or controlled variables) and outputs y_t (measured performance variables) during the early stages of performance degradation.
- Build a CVA state space model based on the obtained training data. Calculate model coefficient matrices as per Eq. (27). Calculate the system states through CVA (i.e. (10)) or CVDA (i.e. (22))
- Estimate system future inputs by looking at production plans or control settings.
- Predict the system behavior $\hat{y}_{t(future)}$ according to the constructed CVA-based state space model (25)–(26). The procedure of predicting system behaviors in the future is actually equivalent to estimating the values of y_t for future time instances. Take y_{t+1} as an example (assuming system status at time $t + 1$ is unknown), one first need to estimate the value of x_{t+1} as per (25), then substitute x_{t+1} into (26), since the only unknown variable in (26) is y_{t+1}, its value can be easily computed.

3.2 Determining the Number of Retained States

Similar to the procedures described in Sect. 2.2, the retained state q (9) is an important tunning parameter that would affect the performance of CVA state space model. Although various approaches have been put forward, for instance, the "knee point" method, determining the dominant singular values and those based on cross-validation. We suggest to use the cross-validation method, and the idea of this approach is to find the optimal q through minimizing prediction error during cross-validation. In order to do so, the training data set need to be divided into two parts, one for constructing the CVA state space model and computing model coefficient matrices; the other for determining the optimal number of retained state q.

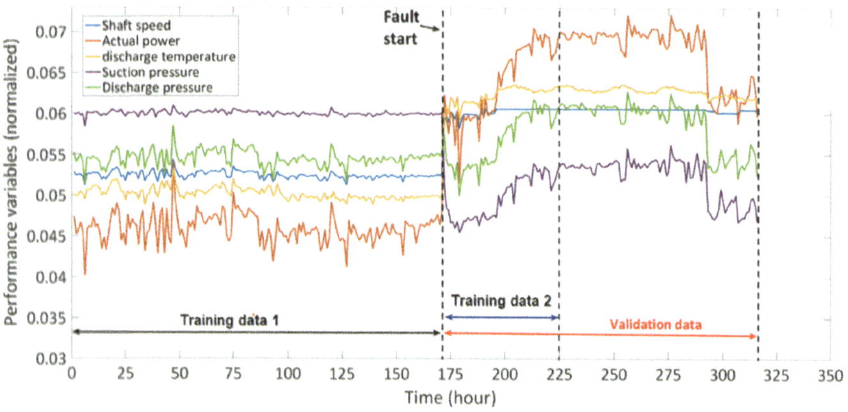

Fig. 5 Sensory measurement for prognosis

3.3 *Example of Using CVA State Space Model for Prognosis*

We demonstrate a case study on a centrifugal compressor. This machine is equipped with various sensors to enable online condition monitoring. As shown in Fig. 5, the machine is in the healthy condition during the first 170 samples of operation. A variable speed drive fault happened afterwards and lasts until the 316 sample. After that time, the malfunctioning drive was removed and replaced with a new drive.

The training data consists of two sets, namely, training data 1 and training data 2 respectively. Training data 1 is adopted to calculate the model coefficient parameters (14), and training data 2 is utilized to determine the number of retained state. The constructed CVA state space model is then validated using the validation data set.

The summed mean absolute and mean absolute percentage error in terms of prediction error over training data 2 are plotted against different numbers of retained states q in Fig. 6 for the determination of q . q was finally set to 1 to obtain the optimal model that provides the highest predictive accuracy.

Figures 7and 8 show two exemplary results in terms of the predicted system behavior over the entire timeframe of degradation. The advantages of CVA state space model is obvious—it is able to track the stochastic fault developments very well. But at the same time, it requires the correlations between system input and output do not change too much during the fault evolution, and the future system input is available in advance.

3.4 *CVA-Based Data Driven Models*

The output of CVA T^2 and T_d can serve as a health indicator of the system, and this indicator is often utilized in combination with predictive data driven models to

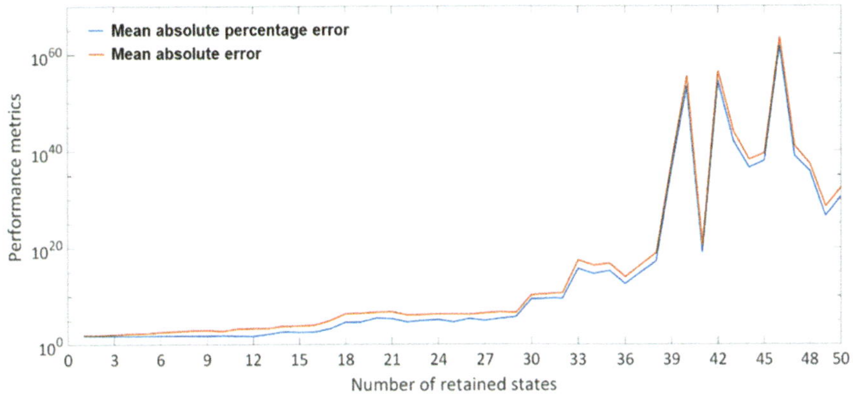

Fig. 6 Determination of retained states

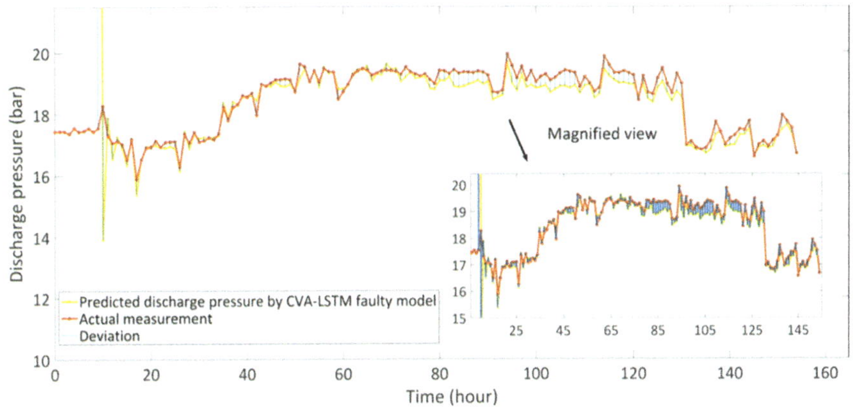

Fig. 7 Predicted system future behavior

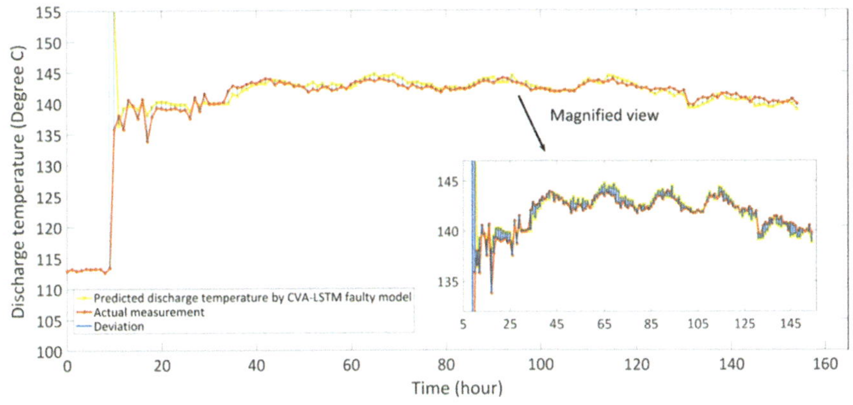

Fig. 8 Predicted system future behavior

form a prognostic scheme. If the calculated health indicator shows a strong degradation trend that can be expressed as, say, an exponential regression model, and simultaneously the historical failure data are scarce. One can predict its future value through projecting time trend. On the other hand, if the health indicator contains a lot of stochastic components, making it difficult to project the time trend, but abundant historical failure data are available, one can consider adopting various machine learning techniques to learn the correlations between these health indicators and their corresponding remaining useful life (RUL). The learnt model can then serve as a predictor to estimate the RUL for a new system.

In this section we demonstrate two examples to illustrate how CVA-based data driven models are applied to carry out prognostic tasks. The role of CVA in these prognostic models lies in reducing the dimensionality of the measured data, thereby facilitating the subsequent data-driven predictive methods.

3.4.1 Example 1—Prognosis via Projecting Time Trend

The first example involves degradation data collected from an industrial centrifugal pump. The measured time series consists of 380 observations and 13 variables. As shown in Fig. 9, the system is operating under healthy condition for the first 334 samples and then had a performance degradation until the end of the time series. The readings of the four different bearing-temperature sensors are captured and illustrated in the figure.

As shown in Fig. 10, we adopted a grey multivariate forecasting model (GMFM) to fit the health indicator (shown with the blue curve in Fig. 10) calculated through CVA. This indicator was computed based on the measurements illustrated in Fig. 9. A small portion of the health indicator at early degradation stages is firstly employed to train a GMFM prediction model. Then GMFM was used in combination with particle filter (PF) to perform time trend projection such that the future value of this health indicator can be estimated with an associated confidence level (GMFM performs the estimation and PF calculates the confidence level). The estimated health indicator is compared with the system failure threshold, and the time at which the

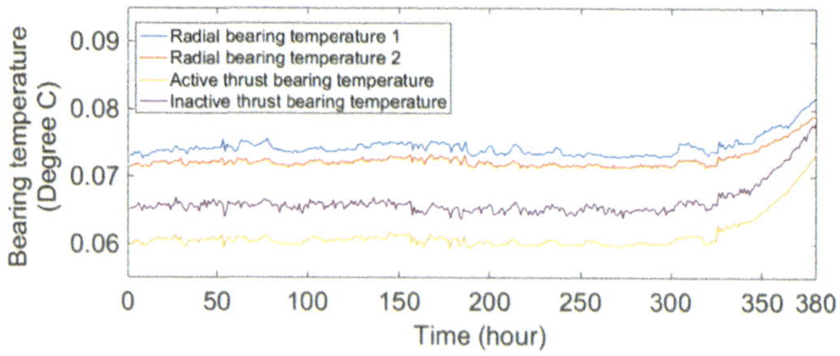

Fig. 9 Measurements for prognostic analysis

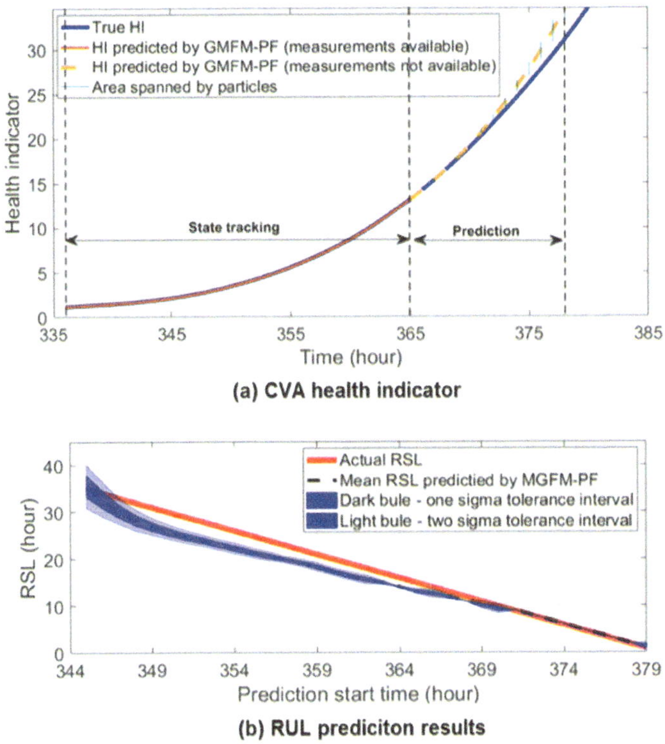

Fig. 10 Prognostic results obtained through projecting time trend

predicted health indicator reaches the threshold are considered as the RUL. The predicted RUL for this machine is demonstrated in Fig. 10 as well. It is observable from Fig. 10 that the predictive accuracy is lower at the beginning and the estimated RULs lie well within the $\mp 25\%$ confidence bounds, indicating that model has the ability to accurately predict the system's remaining service life. As the prediction start point approaches the system failure point, the predicted RUL gets closer to the actual RUL. Overall, Projecting CVA health indicator for RUL prediction is a promising tool for prognostic analysis when the available historical data is scarce. For more details regarding this case study, authors are referred to [19].

3.4.2 Example 2—Prognosis via Machine Learning Approaches

This example involves using a machine learning model, namely gradient boosting tree, to learn the relationship between the various historical CVA degradation indicators and their corresponding RUL values. The learnt model is subsequently adopted to predict the RUL in the presence a new fault.

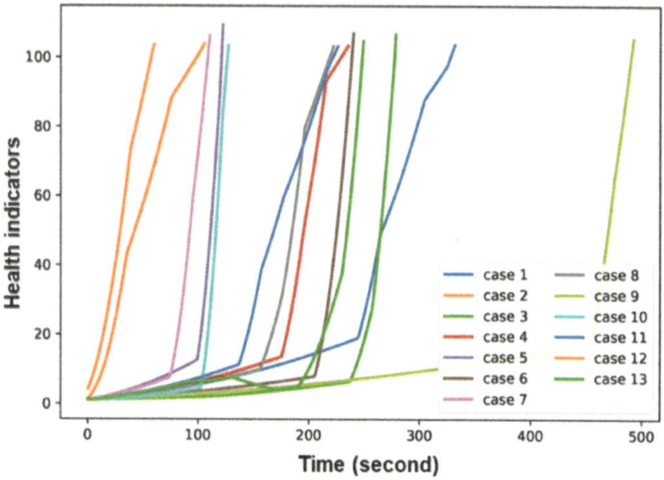

Fig. 11 Measurements for prognostic analysis

The data used in this example was collected from a four-cylinder compressor. This unit experienced thirteen valve failures. The root cause of these failures was found to be improper sealing of the valve due to a missing piece from the outer structure of valve plate. These failures happened at either the head end (HE) or the crank end (CE) discharge valve. Only temperature sensory data were recorded. Eight temperature ratios, namely Suction temperature HE/CE cylinder 1–4 and Discharge temperature HE/CE cylinder 1–4 were utilized by the site engineer for monitoring the health status of the valves. The calculated health indicators for 13 failure cases are illustrated in Fig. 11.

In this case study we developed a machine learning approach called just-in-time-learning (JITL) based gradient boosting decision tree (GBDT) model. The health indicators as illustrated in Fig. 11 are categorized into two groups, namely, one consists of the testing machine and the rest forms the training group. The health indicators in the training group are fed as inputs into the developed JITL-GBDT model, and the corresponding training outputs are the true RULs of these failure cases. The trained JITL-GBDT model is then used to predict the RUL for the testing group. The prognostic results for failure cases 1–12 are illustrated in Fig. 12. We also compare the JITL-GBDT model with the traditional GBDT predictor. The GBDT model apparently resulted in overestimated RUL for cases 3, 5, 6, 7 and 10. It also generated underestimated RUL predictions for case 2. This is mainly due to a rapid degradation taking place within short time which makes the underlying dynamics difficult to be captured by the GBDT model. The JITL models, however, overcome the aforementioned problem, therefore properly captured the degradation patterns in these situations where abrupt rises took place. However, the main aim of this case study is to illustrate how CVA can be used in combination with machine learning

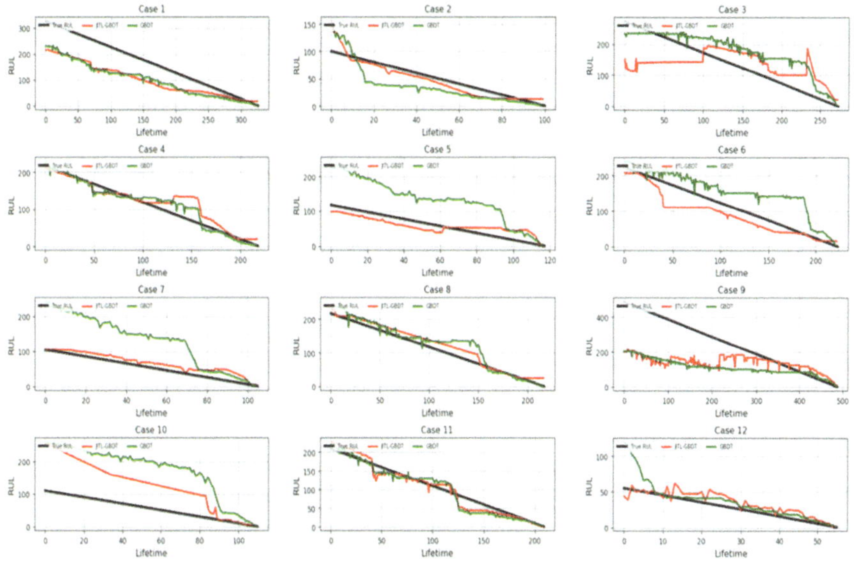

Fig. 12 Prognostic results obtained through machine learning

techniques to predict RUL when historical failure cases are abundant. For more details regarding this case study, authors are referred to [16].

4 Conclusion

This tutorial presented an overview on the methodological evolution of canonical variate analysis with an emphasis on the original linear, canonical variate dissimilarity analysis, CVA-based state space model and CVA-based data driven model. Succinct reviews were also conducted on the adaptive and kernel extensions. The purpose was to explain the theoretical foundations of the variants and to demonstrate how they can be applied to real-world for diagnosis and prognosis through industrial case studies. The applicabilities of the different CVA extensions in relation to the properties of the data were also discussed. The tutorial hopefully can serve as a hands-on tool for applying CVA-based methods in diagnosis and prognosis analysis.

References

1. Adhikari YR (2004) Inference and decision making methods in fault diagnosis system of industrial processes. IFAC Proc Vol (IFAC-PapersOnline) 37(16):193–198. https://doi.org/10.1016/S1474-6670(17)30873-X

2. Al-Najjar B, Wang W (2001) A conceptual model for fault detection and decision making for rolling element bearings in paper mills. J Qual Maint Eng 7(3):192–206. https://doi.org/10.1108/13552510110404503
3. Awadallah MA, Morcos MM (2003) Application of AI tools in fault diagnosis of electrical machines and drives–an overview. IEEE Trans Energy Convers 18(2):245–251. https://doi.org/10.1109/TEC.2003.811739
4. Bach FR, Jordan MI (2003) Kernel independent component analysis. J Mach Learn Res 3(1):1–48. https://doi.org/10.1162/153244303768966085
5. Chen Q, Goulding P, Sandoz D, Wynne R (1998) The application of kernel density estimates to condition monitoring for process industries. In: Proceedings of the American Control Conference, *6* (June), pp 3312–3316. https://doi.org/10.1109/ACC.1998.703187
6. Chen Z, Liu C, Ding S, Peng T, Yang C, Gui W, Shardt Y (2020) A just-in-time-learning aided canonical correlation analysis method for multimode process monitoring and fault detection. IEEE Trans Ind Electron 0046(c):1. https://doi.org/10.1109/tie.2020.2989708
7. Harrou F, Nounou MN, Nounou HN, Madakyaru M (2013) Statistical fault detection using PCA-based GLR hypothesis testing. J Loss Prev Process Ind 26(1):129–139
8. Healy MJR (1957) A rotation method for computing canonical correlations. Math Comput 11(58):83–83. https://doi.org/10.1090/s0025-5718-1957-0085600-6
9. Hotelling H (1936) Relations between two sets of variates. Biometrika 28(3/4):321–377
10. Hyvärinen A (1997) Independent component analysis by minimization of mutual information
11. Jiang B, Huang D, Zhu X, Yang F, Braatz RD (2015) Canonical variate analysis-based contributions for fault identification. J Process Control 26:17–25. https://doi.org/10.1016/j.jprocont.2014.12.001
12. Juricek BC, Seborg DE, Larimore WE (2004) Fault detection using canonical variate analysis. Ind Eng Chem Res 43(2):458–474
13. Li X, Duan F, Loukopoulos P, Bennett I, Mba D (2018) Canonical variable analysis and long short-term memory for fault diagnosis and performance estimation of a centrifugal compressor. Control Eng Pract 72:177–191. https://doi.org/10.1016/j.conengprac.2017.12.006
14. Li X, Yang Y, Bennett I, Mba D (2019) Condition monitoring of rotating machines under time-varying conditions based on adaptive canonical variate analysis. Mech Syst Signal Process 131:348–363. https://doi.org/10.1016/j.ymssp.2019.05.048
15. Li X, Duan F, Loukopoulos P, Bennett I, Mba D (2018) Canonical variable analysis and long short-term memory for fault diagnosis and performance estimation of a centrifugal compressor. Control Eng Practice 72. https://doi.org/10.1016/j.conengprac.2017.12.006
16. Li X, Mba D, Lin T, Yang Y, Loukopoulos P (2021) Just-in-time learning based probabilistic gradient boosting tree for valve failure prognostics. Mech Syst Signal Process 150:107253. https://doi.org/10.1016/j.ymssp.2020.107253
17. Li X, Yang X, Yang Y, Bennett I, Mba D (2019) A novel diagnostic and prognostic framework for incipient fault detection and remaining service life prediction with application to industrial rotating machines. Appl Soft Comput 105564. https://doi.org/10.1016/j.asoc.2019.105564
18. Li X, Yang X, Yang Y, Bennett I, Mba D (2019) A novel diagnostic and prognostic framework for incipient fault detection and remaining service life prediction with application to industrial rotating machines. Appl Soft Comput 82:105564. https://doi.org/10.1016/j.asoc.2019.105564
19. Li X, Yang X, Yang Y, Bennett I, Mba D (2019) An intelligent diagnostic and prognostic framework for large-scale rotating machinery in the presence of scarce failure data. Struct Health Monitor 1–16. https://doi.org/10.1177/1475921719884019
20. Li W, Qin SJ (2001) Consistent dynamic PCA based on errors-in-variables subspace identification. J Process Control 11(6):661–678
21. Negiz A, Cinar A (1997) Statistical monitoring of multivariable dynamic processes with state-space models. AIChE J 43(8):2002–2020. https://doi.org/10.1002/aic.690430810
22. Odiowei PEP, Yi C (2010) Nonlinear dynamic process monitoring using canonical variate analysis and kernel density estimations. IEEE Trans Industr Inf 6(1):36–45. https://doi.org/10.1109/TII.2009.2032654

23. Pilario KES, Cao Y, Shafiee M (2019) Mixed kernel canonical variate dissimilarity analysis for incipient fault monitoring in nonlinear dynamic processes. Comput Chem Eng 123:143–154. https://doi.org/10.1016/j.compchemeng.2018.12.027
24. Pilario KES, Cao Y (2018) Canonical variate dissimilarity analysis for process incipient fault detection. IEEE Trans Ind Inform 3203(c):1–8. https://doi.org/10.1109/TII.2018.2810822
25. Quatrini E, Li X, Mba D, Costantino F (2020) Fault diagnosis of a granulator operating under time-varying conditions using canonical variate analysis. Energies 13(17). https://doi.org/10.3390/en13174427
26. Russell EL, Chiang LH, Braatz RD (2000) Fault detection in industrial processes using canonical variate analysis and dynamic principal component analysis. Chemom Intell Lab Syst 51(1):81–93. https://doi.org/10.1016/S0169-7439(00)00058-7
27. Samuel RT, Cao Y (2015) Kernel canonical variate analysis for nonlinear dynamic process monitoring. IFAC-PapersOnLine 28(8):605–610. https://doi.org/10.1016/j.ifacol.2015.09.034
28. Shang L, Liu J, Zhang Y (2016) Recursive fault detection and identification for time-varying processes. Ind Eng Chem Res 55(46):12149–12160. https://doi.org/10.1021/acs.iecr.6b02653
29. Stefatos G, Hamza AB (2010) Dynamic independent component analysis approach for fault detection and diagnosis. Expert Syst. Appl 37(12):8606–8617
30. Venkatasubramanian V, Rengaswamy R, Kavuri SN (2003) A review of process fault detection and diagnosis part II: qualitative models and search strategies. Comput Chem Eng 27(3):313–326. https://doi.org/10.1016/S0098-1354(02)00161-8
31. Venkatasubramanian V, Rengaswamy R, Kavuri SN, Yin K (2003) A review of process fault detection and diagnosis part III: process history based methods. Comput Chem Eng 27(3):327–346. https://doi.org/10.1016/S0098-1354(02)00162-X
32. Venkatasubramanian V, Rengaswamy R, Yin K, Kavuri SN (2003) A review of process fault detection and diagnosis part I: quantitative model-based methods. Comput Chem Eng 27(3):293–311. https://doi.org/10.1016/S0098-1354(02)00160-6
33. Xu Y, Fan C, Zhu Q, He Y, Hu Q (2019) A novel pattern matching-based fault diagnosis using canonical variate analysis for industrial process. In: Proceedings of 2019 IEEE 8th data driven control and learning systems conference (DDCLS 2019), pp 1132–1136. https://doi.org/10.1109/DDCLS.2019.8909051
34. Yin S, Zhu X, Kaynak O (2014) Improved PLS focused on key-performance-indicator-related fault diagnosis. IEEE Trans Industr Electron 62(3):1651–1658

A Structured Approach to Machine Learning Condition Monitoring

Luca Capelli, Giulia Massaccesi, Jacopo Cavalaglio Camargo Molano, Federico Campo, Davide Borghi, Riccardo Rubini, and Marco Cocconcelli

Abstract The aim of the chapter is to explain the basic concepts of Machine Learning applied to condition monitoring in Industry 4.0. Machine learning is a common term used today in different fields, mainly related to an automated and self-learning routine in a decisional process. This chapter details how a Machine Learning approach may be structured, starting from a distinction between Supervised and Unsupervised approaches. These two classes have different advantages and disadvantages that constrain their application to specific boundary conditions. Machine Learning techniques are the core part of a structured methodology for the condition monitoring, but other phases, such as the pre-processing of data, the feature extraction and the evaluation of performances, are equally important for the success of a condition monitoring system. Together with standard parameters used to assess the performances of the machine learning method, a particular emphasis will be given to the interpretability of the results that can be determinant in the choice and development of a specific tool for condition monitoring in an industrial environment.

L. Capelli · G. Massaccesi · F. Campo · D. Borghi
Tetra Pak Packaging Solutions, Via A. Delfini 1, 41123 Modena, Italy
e-mail: luca.capelli@tetrapak.com

G. Massaccesi
e-mail: giulia.massaccesi@tetrapak.com

F. Campo
e-mail: federico.campo@tetrapak.com

D. Borghi
e-mail: davide.borghi@tetrapak.com

J. C. C. Molano · R. Rubini · M. Cocconcelli (✉)
University of Modena and Reggio Emilia, Via G. Amendola 2, 42122 Reggio Emilia, Italy
e-mail: marco.cocconcelli@unimore.it

J. C. C. Molano
e-mail: jacopo.cavalagliocamargomolano@unimore.it

R. Rubini
e-mail: riccardo.rubini@unimore.it

F. Chaari et al. (eds.), *Smart Monitoring of Rotating Machinery for Industry 4.0*,
Applied Condition Monitoring 19, https://doi.org/10.1007/978-3-030-79519-1_3

1 Introduction

The increasing complexity in automatic machines, combined with the ever-growing need to keep high quality, reliability and efficiency, requires automatic machine manufacturers to develop a new complex model to improve maintenance strategies [1]. New technologies developed in the field of cyber-physical systems (CPS) and Internet of Things (IoT) systems enable these companies to access a large dataset of field data. These technologies manage interconnected physical systems, such as actuators and sensors with cyber-computational capabilities, for example in the case of computer networks, intelligent data management for Big Data and analytical proficiency [2]. Condition Monitoring methodologies, also called Condition Based Maintenance (CBM), allow a real-time diagnostics of the machine health and detect a possible failure of the machine with weeks or even months in advance. This enables a Predictive Maintenance approach. The term Predictive Maintenance has been used for many years and represents one of the business cases enabled by the fourth industrial revolution. In its industrial sense, it concerns the identification of an incipient component or function failure by the use of data collected from the machine, aggregated and processed by means of appropriate algorithms. To be effective, the identification of the fault is to be carried out in good time, i.e. not too late, to allow a reaction (Time to React), but not too early (Useless Detection Window), which would lead to change a substantially still healthy component. If the alarm arrives too late (i.e. there is no time to react) or if the failure causes an unscheduled downtime, there is a Missed Alarm; while, if it arrives too early, there is a False Positive. These indicators are fundamental to establish the effectiveness of analytics used in terms of Confusion Matrix, and more particularly in terms of Accuracy (a general but not very specific indicator), Precision (which measures the goodness of the result compared to False Positives) and Recall (which measures the goodness of the result compared to Missed Alarms). Predictive Maintenance therefore offers the possibility of reducing unscheduled stops and increasing the reliability of the automatic machine. It allows, eventually, to optimize the production and maintenance planning, and finally, up to Prescriptive Maintenance, to advise the customer to carry out appropriate operations and to optimize production plans. For the user, the advantage is to make the programming of his production more predictable over medium and long term, above all avoiding unscheduled stops and reducing maintenance costs. For the manufacturer, the opportunity is to understand and fully characterize the behaviour and the inevitable deterioration of machines, in order to provide value to their customers. This aspect cannot be totally conducted with an accurate Preventive Maintenance, because there is always a statistical queue of deterioration events that occur before the scheduled maintenance event. Further opportunities for both the manufacturer and the user are in turn, to minimize unplanned stops, to discover the root cause of the fault and to reduce the waste due to the equipment failure, with the result of increasing efficiency and sustainability. In the industrial field, this approach also transforms human service work by improving the collaborative human–machine skills for decision-making with respect to maintenance [3]. The collaborative actions between condition-monitoring systems and human service operations involve a socio-cyber-physical system (SCPS) [4]. These systems are linked in a global production network where the interaction of global and individual decision-makers acts in a different way for each sub-system [5]. Nowadays many technologies are available for equipment monitoring, starting from IoT sensors on edge capable of transferring real time data and pre-processing data to cloud computing. The direct result is Big Data Analysis, which refers to the capability of analyzing large datasets collected on the cloud, often using expert systems. In this case, a Machine Learning approach, also known as Data-Driven approach, is very useful because it uses historical data to create a model of the system. The Machine Learning techniques used in CBM can be divided into Supervised and Unsupervised. The former are used when both the sensor data and information about failures are available, while the latter are used only when the sensor data are available [6, 7]. This paper describes a possible approach to develop Machine Learning models, starting from the analysis of the root cause of the problem and the selection of the most appropriate methodologies, and illustrates the main steps and criticalities. The chapter is organized as follows: Sect. 2 gives a general description of Machine Learning and Deep learning; Sect. 3 deals with the steps necessary for the development of Machine Learning (ML) and Deep Learning (DL) models; Sect. 4 describes a general workflow for the ML and DL model creation in Industry 4.0 and Sect. 5 explains the conclusions.

2 Machine Learning

Machine Learning is a branch of Artificial Intelligence (AI) that enables a system to improve its abilities by means of data, while other systems cannot improve their abilities, since they are fixed by hard coded programs. Machine Learning (ML) algorithms are designed to draw elements of knowledge from data and subsequently to apply what they have learned. The most used ML algorithms are based on statistical analysis and data mining. The statistical knowledge is combined with the techno- logical knowledge of the problem to train algorithms in the best way. A Machine Learning model is defined as "the output generated when you train your machine-learning algorithm with data" [8]. In general, as previously mentioned, ML algorithms can be divided into Supervised and Unsupervised. The former are called Supervised because the learning of algorithms is driven and overseen by an "expert teacher", who labels the data and chooses the most significant pre-processed features to be given to algorithms. In this case, it is necessary to know the system behavior previously. In CBM, Supervised techniques translate the competences of experienced technicians into a model that can be used with a large amount of data and in a large number of cases. There are several examples of ML classifiers used for fault detection and fault classification for different equipment [9, 10]. Another example is the use of regressors, which combine multiple signals to predict the future state of a variable concerning the health of the system [11]. In the case of Unsupervised algorithms, input data are not labelled and algorithms are able to automatically find patterns and clusters in order to organize data by similarity. This method is suited to the case in which a large amount of data are unlabelled and consequently a supervision of input information (e.g. target class labels) is not possible in the learning process. Unsupervised algorithms are extremely useful for identifying very complex patterns and their results can be used as input for Supervised algorithms [12].

2.1 Deep Learning

Deep Learning (DL) is a class of ML techniques that uses hierarchical layers of Artificial Neural Network (ANN) to learn from input data. The ANN tries to emulate biological neurons, it consists in three or more layers: an input layer, one or more hidden layers and an output layer. Data enter the input layer, subsequently they are modified in the hidden layers by applying weights and they give the output in the last layer. At this point, the difference between the network output and the expected output is computed and the result is called error. Afterwards the network adjusts the weights of the inner layers till the error rate is reduced. DL can have several hidden layers connected to one another; it depends on the complexity of problems. Deep learning is often used with unstructured data, in this case the features of input data are not pre-computed, but they are directly given as input. For this reason, DL algorithms require a vast amount of data since, in contrast with Supervised techniques, they have to learn by themselves the most suitable features for the task.

A Deep Learning Unsupervised algorithm describes any process that tries to learn a structure and to identify clusters without any identified output or feedback [13]. Deep Learning is based on the fact that the physical problem is not known a-priori. The behavior of the physics of the system is acquired by observing the system itself and it is not taught by an external teacher/knowledgeable expert. That is why Unsupervised techniques are useful when the problem cannot be simply described by a number of variables, operative conditions or features. In CBM a large amount of data are time series coming from sensors [14]. For the analysis of this type of variables, the most used DL algorithms are the ones that handle sequential data, such as Recurrent Neural Network (RNN), Long Short-Term Memory (LSTM) [15] and Gated Recurrent Unite (GRU). Other types, such as Convolutional Neural Network (CNN) [16] and Auto-Encoders (AE) [17] are for feature extraction. In conclusion, Fig. 1 illustrates a very useful comparison of the two techniques, which might help to choose between Machine Learning and Deep Learning. The performances of Machine Learning are asymptotically dependent on the calculated features, the completeness and setting of training dataset. DL leverages the amount and the differentiation of input data. If the two approaches are evaluated in function of the amount of input data, the initial performances of a Machine Learning model are significantly higher than those of a Deep Learning model. With the increase of input data, it is possible to notice that the strict dependence of Machine Learning on the constraints and bounds of human experience limits is asymptotical. Therefore, at a certain point, a Machine Learning model will reach a limit impossible to be overcome because of

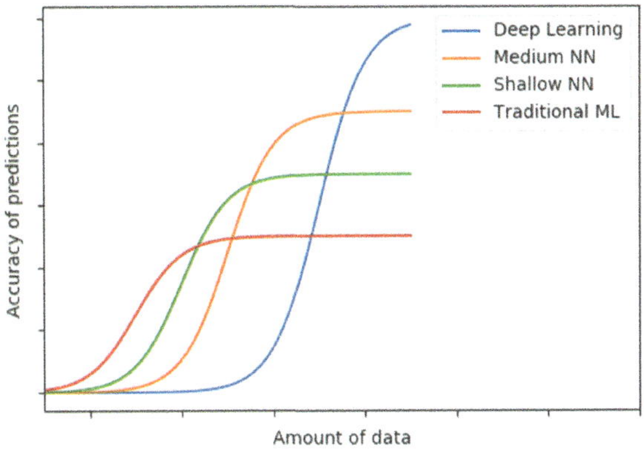

Fig. 1 Performance comparison between Deep Learning (DL) and Machine Learning (ML) [18]

the constraints imposed by both human knowledge and the algorithms themselves. On the contrary, a Deep Learning model is not linked to human experience and will reach and overcome the Machine Learning limits by increasing the quantity of input data.

2.2 Advantages and Drawbacks of the Machine Learning Supervised and Unsupervised Techniques in CBM

As previously mentioned, Machine Learning Supervised and Unsupervised techniques represent the two most common methods, in which the algorithms can automatically learn by experience. They are designed to acquire knowledge from existing data and to use this information for making predictions. Before describing these approaches, it is necessary to specify the differences among "big", "medium", and "small" dataset. The size of a dataset can greatly impact the effectiveness of the model. For a Supervised model, the size of a dataset does not only depend on the number of samples, i.e. the time span of acquisition and the sampling frequency, but also on the kind of approach that the analyst wants to adopt. A practical example can be given by the evaluation of the required size of a dataset for a Supervised algorithm necessary to detect a specific component failure. Supposing that the component under study has a useful life of 1000 h and it is monitored by a vibration sensor, the dataset can be:

- **Small**: if a machine data is one-month sampled once per day and the dataset contains 1 or 2 failure events.
- **Medium**: if data referring to about 5 machines are one-year sampled twice per day and the dataset contains at least 4 or 5 failure events.
- **Big**:if data referring to about 15 machines are sampled three times per day for more than 5 years and the dataset contains approximately 30 or 50 failure events.

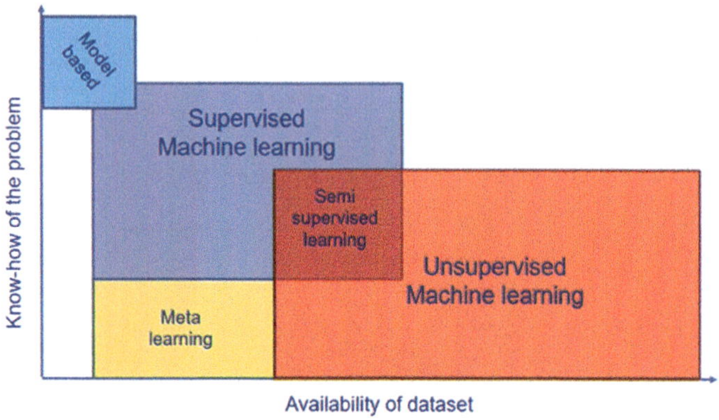

Fig. 2 Comparison and visualization of the different techniques

Differently, as regards Unsupervised techniques, the definition of "big", "medium" or "small" dataset strictly depends on the application and on the model accuracy desired. Every dataset is different, and the number of samples is arbitrary. In this chapter, a general definition is given according to which a dataset can be:

- **Small**: if it includes from 10^2 to about 10^3 samples.
- **Medium**: if it includes from 10^3 to about 50×10^3 samples.
- **Big**: if it includes 50×10^3 samples or more.

Supervised and Unsupervised approaches have different strengths and weaknesses that constrain their application to specific cases. The comparison between the two techniques is described by taking into consideration the five typical different cases that can be found in the industrial field (Fig. 2).

The five different scenarios are the following:

- **Big dataset of unlabelled available data with a low know-how of the equipment**. This circumstance can happen when new pieces of equipment are placed on the field and operators have not a great experience with possible future failures. If there is a quite large dataset that consists in signals coming from different sensors, a good approach could be the use of Unsupervised method. De facto, in a situation in which the conditions of the monitored equipment are not well definable and limited, a Machine Learning Supervised approach would be too strictly human-dependent: its use could cause the problem to be bound by the mind-set of the expert who trains the model. A good choice is the creation of an anomaly detector. The anomaly detector is a class of Unsupervised algorithms that allows to detect the behaviour variation of a system without any label. This kind of algorithms is based on a statistical analysis of the signals by which a "normal" status of the system is defined. When there are drifts or outliers from the normal behaviour, an alarm is produced. This method has been used in several cases [19–21].

- **Medium dataset of well-known equipment**. This is the case in which the quality of feedback is well defined; the labelling process is better performed; the state of the problems in faulty cases is well defined by specific classes; data coming from the machinery under exam are sampled in different working conditions, they are well balanced and explanatory of the problems. In this situation, the most suitable approach is the Machine Learning Supervised. It allows to distinguish different classes of failures and it can be very specific in class definition. In this way, the classifier is trained to have an accurate decision boundary in distinguishing different classes. Knowledgeable experts play a key role in the training of Supervised Learning models in manufacturing applications. With their experience, they can direct the algorithms towards the main classes that can be detected through some parameters. They well know how the system recognizes some specific malfunctioning and react to it. In this way, the model can apply this knowledge and generalize it. A great advantage of Machine Learning is to be able to achieve good performances with fewer training data than Deep Learning, thanks to the useful examples identified by experts. An Unsupervised approach is not suggested due to the dataset size and the human knowledge that can accelerate the training of Supervised algorithms. The Supervised algorithms are usually faster, simpler and surprising flexible, if data are well pre-processed and the dataset contains useful engineered features. De facto, the main characteristic of the Supervised Machine Learning techniques is the ability to handle large amounts of high-dimensionality data quickly and reliably, with a small computational cost. Another important advantage provided by a Supervised Model it is the possibility to have a high interpretability of the results that help the validation of the model.
- **Big amount of data with an incomplete knowledge of the machine**. This situation is the most typical in industry. It is quite difficult to have all the historical data of the equipment to properly know its past abnormal behaviors or maintenance operations performed to correct a faulty condition. In this case, two possible approaches are suggested: Semi-Supervised Learning approach and Unsupervised Learning approach. Semi-Supervised Learning is a hybrid approach in which a trained expert should go through a small subset of a dataset labelling it. It would require too much time and it would be an intense and costly operation, but, in the presence of half-labelled dataset, it is possible to use Deep Learning algorithms for feature extraction with CNN or AE. Semi-Supervised Learning is a branch of Machine Learning that aims to combine the tasks of Unsupervised and Supervised Learning [22], attempting to improve their performances. It is used just in those cases in which a combined dataset of a small amount of labelled data and a large amount of unlabelled data is available. For the classification model, a Semi-Supervised Learning algorithm classifies and assigns a class to those additional data points for which the labels are unknown. Labelled data guide the model for classification, while unlabelled data can help in the construction of a better classifier algorithm with further information. On the other hand, for clustering methods, the learning procedure can benefit from the knowledge that certain data belong to a specific class. As most Machine Learning models, a Semi-Supervised approach is focused on classification [23]. As regards Unsupervised Learning, it

is suggested to apply an anomaly detection algorithm, which identifies extreme points, exceptional events and observations that raise suspicions since they significantly differ from the greater part of the data. Typically, these extreme points are abnormal data, which can be connected to the problems due to some failures or anomalous processes. Therefore, once data are distinguished into two classes "normal" or "anomalous" by the anomaly detector, experts can use the know-how of the machine in order to understand which class includes real failures and which class includes only normal variations of the system. After that, a new dataset is created by means of new labels, which are identified by the anomaly detector and checked by experts. The new dataset is used as input to a classifier to automatize the classification process.

- **Small dataset of data available with a low know-how of equipment**. Normally, with the application of Machine Learning or Deep Learning algorithms, a high amount of data (Big Data) is required to get a robust result. On the contrary, human brain can, occasionally, learn from small dataset in an efficient way. A branch of AI, called Meta-Learning, is scouting this possibility with encouraging results. Basically, the aim of this approach is to develop, after a brief learning session, a model that can adapt to situations never seen before in a robust way. In some respect, Meta-Learning is a way to teach the algorithm to "learn how to learn" and apply it thoroughly. An alternative to these techniques is the use of Data Augmentation approaches, which artificially create, from available data, a larger dataset, in order to fall into one of the previous categories. This extension of the dataset is achieved by making new data, with additional slight changes. For example, for image data, the pictures can be replicated and rotated and/or zoomed [24].
- **Small dataset of data available with a high know-how of equipment**. When a small amount of data is available, but the equipment or process is well or deeply known, a Model Based Machine Learning can be applied. This approach offers the opportunity to develop tailored solutions for specific scenarios, in a white-box manner, enabling effective comparisons of trails and errors among different alternative models. In this technique the focus is on the model itself, for example on engineering and domain knowledge aspects rather than on AI methods.

A practical example to understand how to implement the proper approach in a real environment can be find in the chapter 4 titled *A Structured Approach to Machine Learning for Condition Monitoring: A Case Study*, in this same book [25].

3 Development of Classifiers with Machine Learning Algorithms

In this section all the steps necessary to develop a ML model are described. The first step is data collection, this step is crucial because the quantity and quality of the data collected determine the accuracy of the model. After that, it is necessary to label each branch of data, a dataset is labelled when every instance is associated with a class. As regards Condition Monitoring, data labelling starts with the listing of faulty cases. The aim is to find the period when an issue takes place and which components, functions and failure modes are involved. In this phase, any information is retrieved from the field. Knowing the dates when faults take place and the dates when the components are substituted is critical. These dates determine the threshold within which a certain component can be considered healthy or faulty. The wrong labelling can negatively affect the training of the model. This step is also necessary to classify and cluster the main failure modes of the components and functions of the system. This analysis could be very useful for the decision on how many classes are to be predicted by the model. The main types of classification task are:

- **Binary Classification**. This is the task of classifying the dataset into two classes. In CBM, these two classes usually refer to a normal and abnormal state. The binary classification is applied to those cases in which it is not necessary to predict the type of failure of a certain component, but it is only necessary to predict the state of that component: if it is "Healthy" or "Faulty".
- **Multi-Class Classification**. Unlike Binary Classification, the Multi-class Classification can describe the evolution of a component during its life and the different failure modes. Therefore, classes can vary from "Running-in" or "Healthy" to "Broken" or "Faulty" if the aim is to predict the degradation phase of the system. On the contrary, if the aim is to predict the specific type of damage of a component, classes can assume the "Healthy" label and a different label for each known failure mode.
- **Imbalanced Classification**. It is a classification method in which the classes of the training dataset are unequally distributed. Typically, Imbalanced Classification is a Binary Classification task in which the majority of examples in the dataset belong to the normal class and the minority to the abnormal class. This situation is very common in the early stage of the analysis of a new equipment. This problem requires specialized techniques to change the composition of samples in the training dataset by under-sampling the majority class or over-sampling the minority class (Evolutionary Undersampling) [26], Synthetic Minority Oversampling Technique (SMOTE) [27]).

The next step is data pre-processing. This is the most delicate and critical step in a Machine Learning process. It can be split into the three following stages:

- **Raw data cleaning**. In this first phase, raw data sampled from the equipment under exam are cleaned of several types of problems, such as noise, redundancy of data and missing data. The sensors, the cables or the recording system used might be broken and consequently the recorded signals are not consistent. To detect possible failures of one of the recording system components, it is suggested to use algorithms that can eliminate the inconsistent data automatically. Another type of bad data-recording might be caused by a wrong logging policy of acquisition in which, for instance, the data sampled are not acquired during the desired phase of work of the machine. A wrong logging policy or a logging issue can lead to a redundancy of data or a lack of data acquired during the process. During the

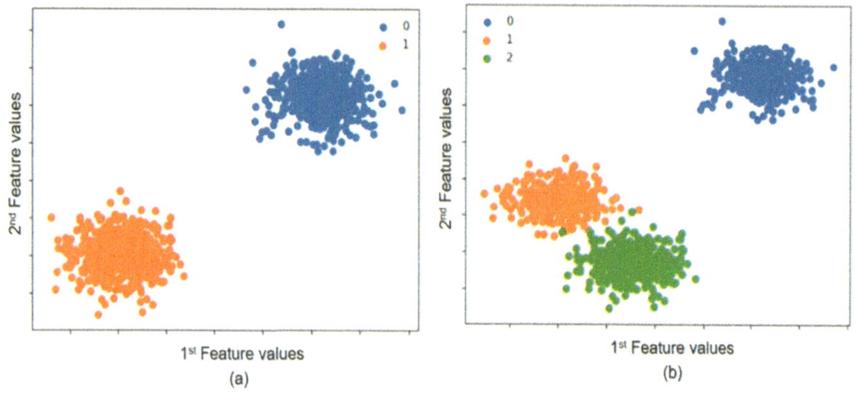

Fig. 3 Scatter plot of: **a** binary classification dataset, **b** multi-Class classification dataset

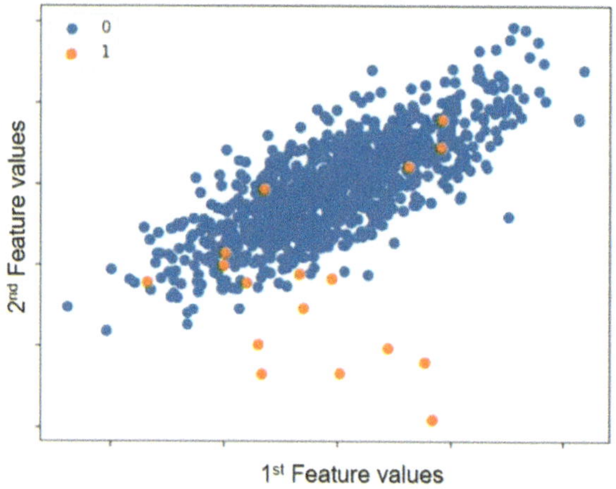

Fig. 4 Imbalanced binary classification dataset

cleaning phase, all the cases listed above will be excluded from the list of raw data that will be used to extract features [28] (Figs. 3 and 4).

- **Feature extraction**. Basic domain knowledge and contextual understanding are the key to create a structured and high-quality dataset. The result of this step is a dataset that consists in a collection of features representing examples of different status of the system. If there is an incomplete knowledge of the features to be used, DL algorithms can be of help to feature extraction, but only for large datasets. Another method consists in starting the analysis by computing several features and subsequently in reducing their number by means of statistical methods, such as Principal Component Analysis (PCA), Linear Discriminant Analysis (LDA), or the use of ML algorithms that weight input features, for example Random Forest [29–32].
- **Dataset cleaning**. The dataset obtained from data pre-processing needs to be cleaned before being given as input to Machine Learning Supervised algorithms.

To clean-up the dataset, it is necessary to handle missing values. An improper data collection could cause a lack of data in the dataset that can be filled with feature values obtained by interpolation, but this solution could add variance to the dataset. The dataset often contains features with different unit of measure, magnitude and range. This can pollute the output of the model. In this case, feature scaling is to be applied. Feature scaling is a technique used to standardize features in a fixed range. There are many ways to scale feature values, the most important are standardization and normalization. Standardization modifies the features in such a way that their mean is equal to 0 and standard deviation to 1, while normalization scales the features between 0 and 1. Feature scaling is often optional and not required, but sometimes it really influences the Machine Learning models [33, 34].

After the data pre-processing, it is a good practice to visualize the data to search possible relationships among features or variables and to check if the dataset is balanced or unbalanced. The balance of the dataset is an important point because, if the ML model is trained with many data belonging to one class, the model can be subject to bias. For example, if the input data have only two classes, one of which containing 90% of input data and the other one containing 10% of input data, the ML model will be biased towards guessing that the larger part of the system status belongs to the first class. But it is not true because the dataset is a small representation of the global status of a system. For this reason, it is important to balance the dataset as much as possible in order that all the classes are balanced. The ML model usually needs three different datasets to be developed: train, test, and validation. The first is used to train the algorithm; the second to test the output of the ML model and to optimize it; the third is used to validate the real performances of the model. An exception to the ML models is the Bayesian Neural Networks that do not necessarily require a validation set. The three groups of data (train, test, validation) are to be balanced and a good practice is to randomize the data order to exclude any bias produced by a time order of the recorded data. The general rule is to use 70 % of the dataset as training data, 20 % as validation data (in order to adjust parameters with the aim of tuning the model) and 10 % as test data (in order to evaluate the accuracy, recall and precision of the final model). The next step is the choice and training of the model. Several ML models are used for CMB. Some of them are well suited to classification (Support Vector Machine [35], Random Forest [36]), other ones to regression (Neural Network regression [37], LASSO regression [38]), while other ML models are suitable for time series analysis (RNN, LSTM). The important thing is to choose a model suited to the task. Subsequently, it is possible to train the model; each model has its own training time; DL models could need a dedicated hardware such as GPUs. After the training, the performances of the model are evaluated through a validation dataset containing data never used in the training process. Validation represents how the model might perform in the real environment. Different tools are used for validation. A Confusion Matrix is a table that allows to measure the performances of a classification algorithm. Each column contains the actual class, while each row represents the instances in a predicted class. Four performance metrics are to be considered: Accuracy, Precision, Recall and F1

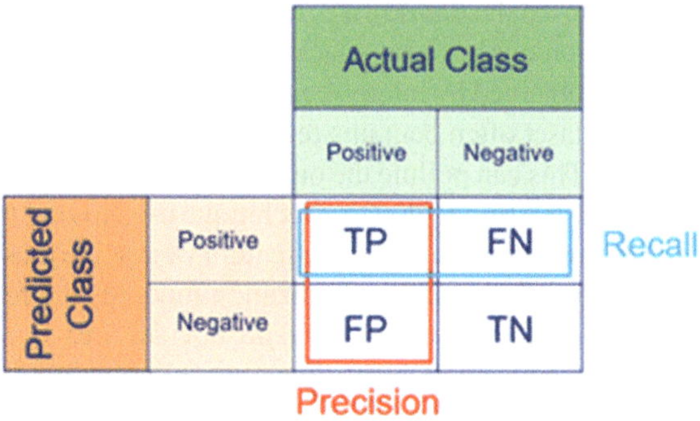

Fig. 5 Confusion matrix nomenclature

Score. Before defining each of them, it is important to define four values related to a general classification:

- **True Positive** (TP): number of correct predictions of Positive class.
- **True Negative** (TN): number of correct predictions of Negative class.
- **False Positive** (FP): number of incorrect predictions of Positive class.
- **False Negative** (FN): number of incorrect predictions of Negative class (Fig. 5).

Here is the description of the four performance metrics:

- **Accuracy**: it measures the overall accuracy of the model classification. It is the ratio between the number of correct predictions and the number of predictions.

$$ACCURACY = \frac{TP + TN}{TP + TN + FP + FN} \tag{1}$$

- **Precision**: it measures the accuracy of a class. It specifies how good the algorithm is for the identification of a Positive class without any false alarms. It is the ratio between the correct predictions of a class and the sum of the correct and incorrect predictions of the same class.

$$PRECISION = \frac{TP}{TP + FP} \tag{2}$$

- **Sensitivity or Recall**: it measures the portion of elements belonging to a class classified in a correct way. It specifies how good the algorithm is for the identification of a Positive class without any false alarms. It is the ratio between the correct predictions of a class and the sum of correct predictions of the same class and incorrect predictions of the opposite class.

$$RECALL = \frac{TP}{TP + FN} \tag{3}$$

- **F1 Score**: it is very common to have an excellent Precision with a bad Recall and vice versa. F1 Score provides a way to express both metrics with a single score.

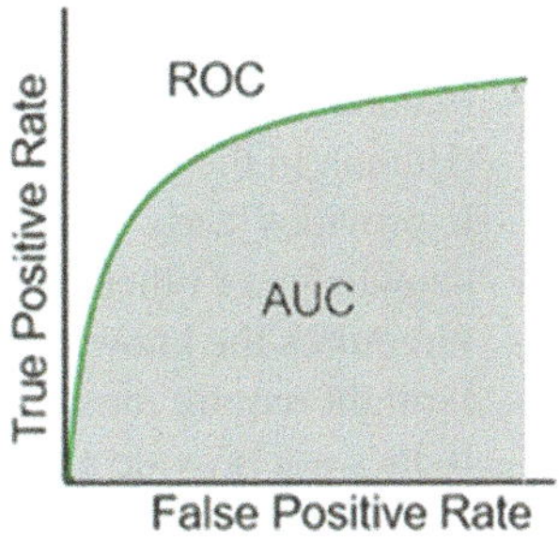

Fig. 6 Receiver operating characteristics (ROC) curve (green line) and area under the curve (AUC) (grey area)

This measure is the variant that is very often used when data are unbalanced. It is calculated as the harmonic mean of Precision and Recall.

$$F1 = \frac{2}{\frac{1}{RECALL} + \frac{1}{PRECISION}} = 2\frac{PRECISION * RECALL}{PRECISION + RECALL} \quad (4)$$

In order to have a graphical representation of the metrics, the ROC (Receiver Operating Characteristics) curve (Fig. 6) is used. It represents the TP and the FP in the same graph. The AUC (Area Under the Curve), which is the area under the ROC curve, is a good estimator of the quality of the model and represents the degree of separability between the classes [39].

After the evaluation, the next step is the parameter tuning. The data expert can optimize the model and test new configurations in order to be more confident about the chosen model and the results. A possible action is to conduct more tests on the model performances by means of different training tests and validation datasets. In this case, if the performances are nearly similar, it means that the dataset has been balanced and randomized properly. The model can also have a problem of overfitting; in this case, it succeeds in well predicting only some classes because of a problem of unbalanced dataset, i.e. the model only recognize itself, and it is not able to generalize the detection on new data. Another action is to change some parameters of the model. In parameter tuning, one of the most used variables is learning rate. It defines the step size used for each learning iteration while the model is minimizing the model error (loss function). This parameter is fundamental for the accuracy of the model and for the time used in the training phase.

4 Model Development Workflow

This section describes the way in which ML models can be implemented and optimized for CBM in Industry 4.0. Figure 7 represents the main approaches to predict maintenance problems and they are not exclusive. The main aspects that characterize each development phase can be summarized as following:

- **Deviation Monitoring**:
  - Manual or semi-automatic approach where human contribution is often required (Human in the loop).
  - It provides early warnings regarding potential failures with or without any knowledge of other similar failures.
  - It requires the knowledge of the system to label its "healthy" conditions.
  - It might require the knowledge of admissible working conditions.
  - It does not provide any predictions or estimation of the dataset.
- **Automatic detection and classification**:
  - Fully automatic (without Human in the loop) detection of the root causes of the failure.
  - Only modelled failures can be detected and classified.
  - It provides additional information on which type of failure will occur and a confidence level of prediction for the modelized failures.
  - Model performances can be measured with precise metrics. Business logics can smooth false predictions.
  - Labelled data are necessary, the model performances are strictly connected to the quality and quantity of the labelled data.
- **Remaining Useful Life**:
  - Automatic (without Human in the loop) estimation of remaining days or cycles before the failure.
  - The estimation is based on observations and modelling of past failures.
  - It is used for long term maintenance of assets.
  - Labelled data are necessary, the model performances are strictly connected to the quality and quantity of the labelled data.

The correct approach can be chosen case by case and it depends on the business case because of the different effort required. Moreover, the type of point of interest, quantity and quality of available data and labels contributes to highlight the proper strategy according to the requirements of each step.

Deviation monitoring phase can provide early warnings or advise about potential issues. It is usually the first step of analytics in CBM. This phase is mainly focused on plotting raw data or pre-processed features with the main purpose of detecting anomalies in the behaviour of the system. This is the most general and easiest approach to be implemented. It requires only the knowledge of the system under analysis without any need to gain access to big amounts of historical data and labels. Limited information on the type of failures and time to failure is usually provided. This approach implies the human expert in the loop; the expert leverages on the available information and the knowledge on the system to determine if the actual condition of the machine is anomalous. This is one of the most common approach CBM for manufacturing lines especially in those cases where the machines under monitoring are different from one another or have different working conditions. Deviation monitoring phase is also useful for defining indicators used in the next phases,

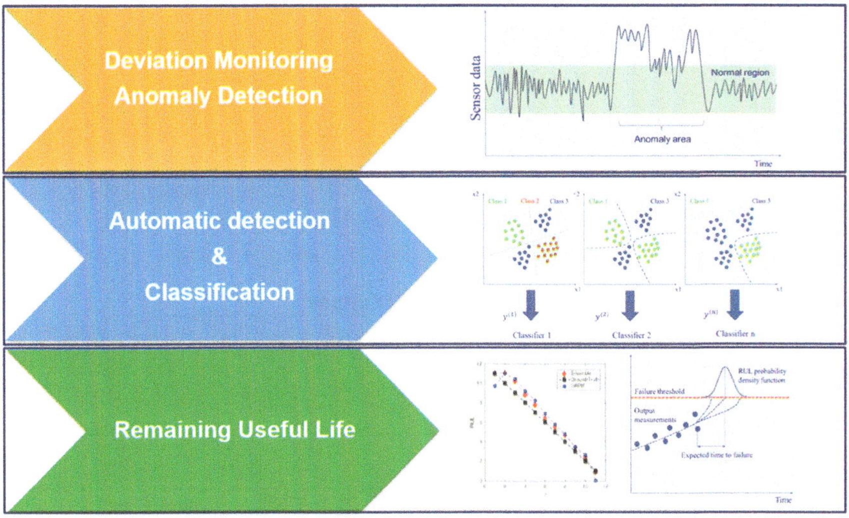

Fig. 7 Condition monitoring landscapes

in particular the observation of the selected indicators leads to confirm their usability or to narrow down the available information. Furthermore, the observation leads to a better understanding of the behaviour of the indicators and the functions with respect to failures and to a redefinition of the business case. As regards rotating machines, for instance, it is important to select the correct indicator on the basis of the specific failure mode. Some indicators, such as impulsiveness and energy content of recorded vibration signals, are very useful and intuitive in the detection of anomalies based on historical data observation. In case of failure classification, these indicators could be insufficient as they represent only two characteristics of the vibration signal. An example is represented in Fig. 8. In this case, the trend of the Root Mean Square (RMS) of the vibration, that represents the energy content of the signal, is constantly increasing with the evolution of the failure over time; while kurtosis, that represents the impulsiveness of the signal, is not monotonically increasing, but it shows sporadic peaks some weeks or months before the failure. The behaviour of these indicators can vary a lot depending on the failure mode.

Typically, in the case of servo motors, these indicators fluctuate during the lifetime, decreasing to a working level after the initial running-in till a failure takes place for the damage of the mechanical components. The impulse indicators remain high just for the short time the failure is generated to come back to a normal or even subnormal level very rapidly. The indicators can sometimes drop below the normal level when the energy content and the general vibration level increase. The loss of rigidity in the system can cause a higher vibration level. In particular, this affects those components under cyclo-stationary state, which generate high levels of kurtosis even during the normal working conditions. The increase in noise can sometimes cover the peaks of the signal normally generated by the cyclo-stationary effect. This makes necessary

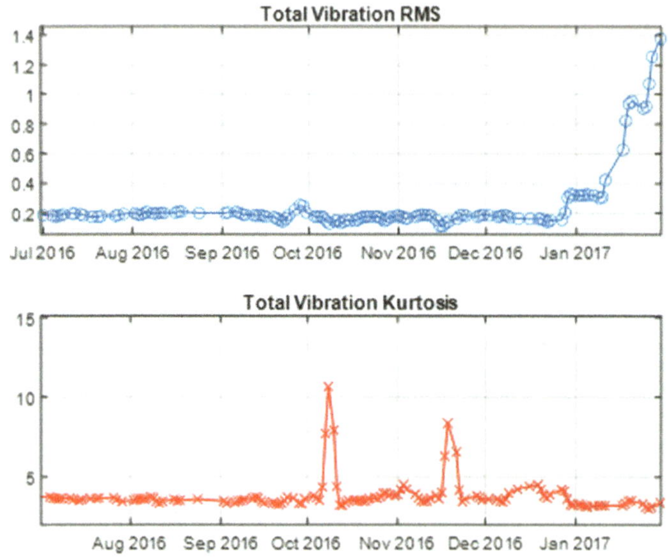

Fig. 8 Indicator behaviour in deviation monitoring phase

a continuous observation of the indicators over time in order to recognize the actual status of the system. For this reason, a proper observation of the indicators and choice are crucial for the next automatic detection phase. The main parameters used for deviation monitoring in rotating machines are the following:

- RMS: it is defined as the square root of mean square;
- Variance: it is the second central moment of a real-valued random variable;
- Skewness: it is the third central moment of a real-valued random variable;
- Kurtosis: it is the fourth central moment of a real-valued random variable;
- Quartiles: they are the 25th, the 50th and the 75th percentiles of the input variable.

Moreover, parameters related to frequency are also used to trend specific indicators over time since they are more connected with failures. They are calculated starting from the kinematical model of the system and considering all the connected components, such as gearboxes, motors, valves etc. The information coming from the model is merged with the information coming from the used sensors, such as encoders, current sensors and accelerometers. The indicators are calculated on the basis of the acceleration or velocity (acceleration integral), of the spectrum of the vibration by computing the energy content (RMS) and the peaks of a specific bandwidth corresponding to the characteristics of the failure.

The typical parameters considered are showed in the failure mode matrix in Figs. 9 and 10 The figures illustrate the theoretical importance of the specific spectral components in relation to the type of failure, respectively for ball bearing components in Fig. 9 and Gearboxes in Fig. 10. For each failure mode described in rows, a specific weight is given to the more probable feature calculated in frequency domain.

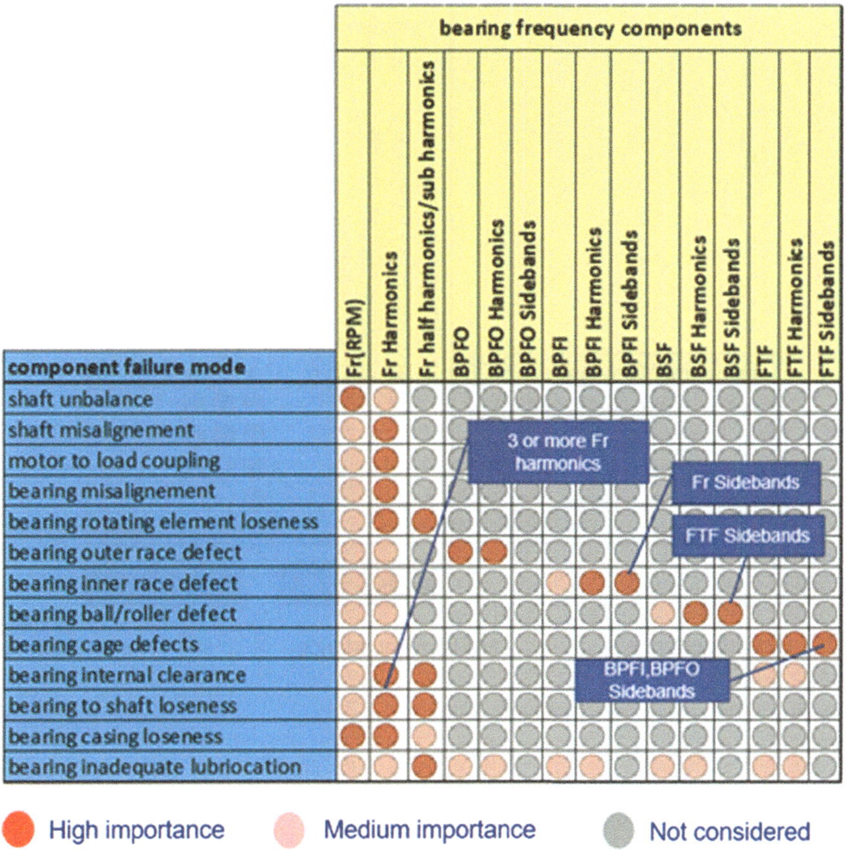

Fig. 9 Ball bearing indicators/failure mode matrix

In this way, it is possible to distinguish and classify failures with similar frequency component behaviour. For example, the energy content of the described frequency bandwidth and its normalized variation over time is computed and visualized in a matrix to cross-correlate the variations with the theorical features behaviour.

Failure classification phase provides outcome on the status of the specific point of interest by warning if the data are classified as potential failures. This phase is focused on the development of all the possible procedures useful for automatically recognizing when the system status changes, but it does not give any information about the estimated failure time. Alerts can be generated on a binary classification or a multiclass classification depending on the failure modes of the components under observation and on the business case. As regards the binary classification, it does not provide any indications of specific failures, but it only warns about anomalies in the conditions of the components. For manufacturing machines, the most appropriate model is always a balance among benefits, simplicity and implementation effort.

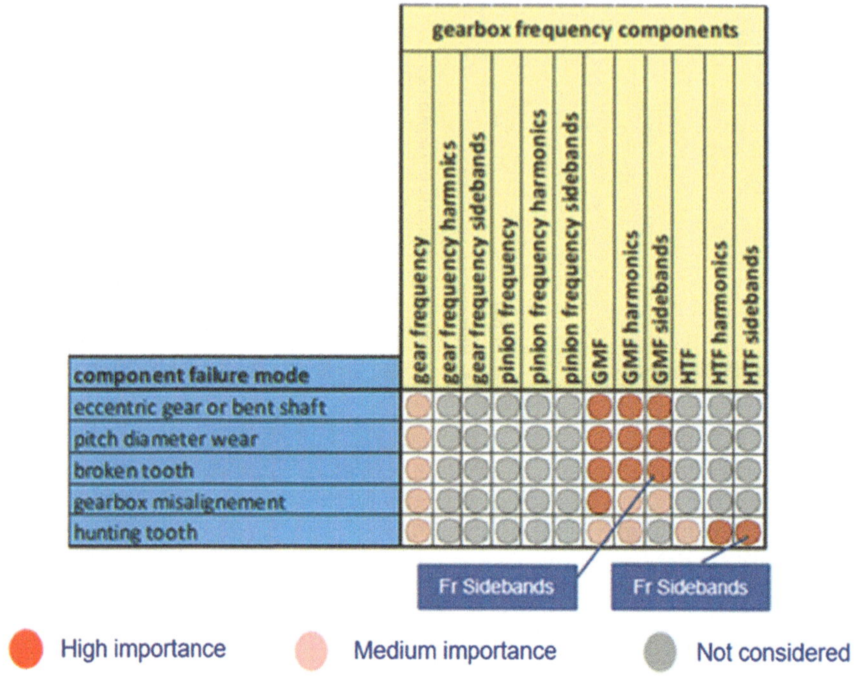

Fig. 10 Gearbox indicators/failure mode matrix

While the multiclass classification is usually done through Machine Learning techniques, the binary classification can be obtained even through a simple threshold control. The multiclass classification can instead offer the possibility of distinguishing several types of failure, but it also requires a bigger and more detailed dataset. Data of several failures are to be available for each failure mode so that it is possible to accurately describe and model the behaviour of monitored failures. Accuracy of the model in respect to all possible failure modes of that function is proportional to the number of similar events present in the dataset; unknown failures cannot be modelized and this decreases the recall generating missed alarms. The main goal of this phase is to automatize the human experience in detecting anomalies through the visualization of selected feature variations. Experts are still in the validation loop to confirm a possible prediction and reinforce the learning of deployed algorithms. Furthermore, depending on the dataset balance, it is possible to train algorithms in a unique class. This is useful for detecting deviations from a "normal" path of the data and subsequently alerting in case of anomalies. Remaining Useful Life phase provides insights in the remaining life of the component or functions useful for scheduling maintenance interventions and optimizing time and resources. Moreover, it is suitable for long term asset maintenance. The application of these techniques make sense only in the case of failures in which time constant between the possible detection and the failure is considerably high and smoothed enough to be predicted. RUL implies a deep knowledge of every failure mode of a specific component and a huge amount of data of failures to be analysed. In this phase it is crucial the availability of several component life cycles (from the asset installation to the asset failure) to characterize the behaviour of the indicators representative of the system status. This is mainly applied to series of production machines where a big amount of cycles for each failure mode can be collected, but it is not sustainable in case of different types of machines or different working conditions. RUL estimation and knowledge of failure modes are the key steps for prescriptive analytics.

There are different methods and techniques [25] that can be listed for RUL estimation. The methods can be grouped into the following categories:

- Model Based: applying statistical and computational intelligence.
- Analytical Based: applying physics based modelling, possibly based on or validated by experimental results.
- Knowledge Based: using a collection of information from domain knowledge experts and interpretating it with computational intelligence.
- Hybrid: using domain knowledge based models and data-driven approaches, in order to improve accuracy.

The techniques, which can be effectively applied for RUL estimation, are elements of above-mentioned methods and can be clustered in the following way:

- Statistics: it is the use of techniques based on statistical analysis, such as Statistical Process Control.
- Experience: it is based on domain knowledge expert judgement; it identifies features giving information about the degradation of mechanism or process and can, in turn, facilitate the preparation of the RUL formula.
- Computational Intelligence: it includes Artificial Neural Networks, Fuzzy Logics, Bayesian techniques, Support Vector Machines, among other techniques.
- Physics of failure: it relies on parametric data and techniques to characterize the failure behaviour and evolution over time.
- Fusion: it applies a merge of datasets of different origin.

5 Conclusions

This chapter presents a workflow to select and optimize a proper technique for Machine Learning in condition monitoring applications. It describes the series of phases necessary for selecting the best methodology according to the problem under exam, the points of interest, the availability and consistency of the dataset. In industry, the increasing mechanical complexity of systems and the big diversification of working conditions require the usage of advanced techniques to detect and classify anomalies, to optimize maintenance events and reduce unplanned stops. The main differences between Supervised and Unsupervised techniques are illustrated and contextualized in typical industrial applications, taking into consideration the most relevant aspects of failures, data availability and result interpretability. Furthermore, the main metrics to measure the performances of the model are explained within the industrial environment, where precision and recall metrics show undetected failures or false alarms that can lead to additional costs for customers. In CBM for industry 4.0, experts' knowledge is combined with powerful statistical tools. In this mixed approach, a crucial phase is the extraction of the proper features based on the model of the system and observability of the failure modes. Thanks to this approach, it is possible to construct a reliable and trustworthy dataset that realistically describes the behaviour of the system with a good degree of generality. Deep Learning techniques are instead a powerful tool that can be represented as a black box where input and output data are known, while the features useful for describing the conditions of the system are unknown. This approach can be used when the phenomena to be described are unknown or they cannot be represented easily. The feature extraction, which usually plays a key role in the description and interpretability of the system, is automatically done by the models on the basis of the input dataset that needs to be larger than the one used in the Machine Learning case. The fundamental pillar in the use of ML and DL in industrial environment is the large-scale utilization of intelligent sensors and IoT devices that allows to collect a large amount of data. The availability of big datasets, together with the refinement and discovery of new algorithms, cannot leave the deep knowledge of the phenomena under observation and the system characterization out of consideration.

References

1. Peng Y, Dong M, Zuo MJ (2010) Current status of machine prognostics in condition-based maintenance: a review. Int J Adv Manuf Technol 50(1–4):297–313
2. Lee J, Bagheri B, Kao H-A (2015) A cyber-physical systems architecture for industry 4.0-based manufacturing systems. Manuf Lett 3:18–23
3. Mobley RK (2002) An introduction to predictive maintenance. Elsevier, Amsterdam
4. Fleischmann H, Kohl J, Franke J (2016) A modular architecture for the design of condition monitoring processes. Procedia CIRP 57:410–415
5. Frazzon EM, Hartmann J, Makuschewitz T, Scholz-Reiter B (2013) Towards socio-cyber-physical systems in production networks. Procedia Cirp 7:49–54
6. Wang K (2016) Intelligent predictive maintenance (ipdm) system–industry 4.0 scenario. WIT Trans Eng Sci 113:259–268
7. Susto GA, Schirru A, Pampuri S, McLoone S, Beghi A (2015) Machine learning for predictive maintenance: A multiple classifier approach. IEEE Transactions on Industrial Informatics 11(3):812–820
8. IBM, https://www.ibm.com/uk-en/analytics/machine-learning?p1
9. Bode G, Thul S, Baranski M, Muller D (2020) Real-world application of machine-learning-based fault detection trained with experimental data. Energy 198
10. Said M, Abdellafou KB, Taouali O (2010) Machine learning technique for data-driven fault detection of nonlinear processes. J Intell Manuf 31:865–884
11. Abdusamad KB, Gao DW, Muljadi E (2013) A condition monitoring system for wind turbine generator temperature by applying multiple linear regression model. In: North American power symposium (NAPS), Manhattan, KS, pp 1–8

12. Wuest T, Weimer D, Irgens C, Thoben K-D (2016) Machine learning in manufacturing: advantages, challenges, and applications. Production & Manufacturing Research 4(1):23–45
13. Deng L, Yu D (2014) Deep learning: methods and applications. Foundations and trends in signal processing 7(3–4):197–387
14. Khan S, Yairi T (2018) A review on the application of deep learning in system health management. Mechanical Systems and Signal Processing 107:241–265
15. Zhao R, Wang J, Yan R, Mao K (2013) A condition monitoring system for wind turbine generator temperature by applying multiple linear regression model. In: North American power symposium (NAPS), Manhattan, KS, pp 1–8
16. Wang F, Dun B, Liu X, Xue Y, Li H, Han Q (2018) An enhancement deep feature extraction method for bearing fault diagnosis based on kernel function and autoencoder. Shock Vibr 6024874
17. Sun W, Zhao R, Yan R, Shao S, Chen X (2017) Convolutional Discriminative Feature Learning for Induction Motor Fault Diagnosis. IEEE Transactions on Industrial Informatics 13(3):1350–1359
18. Del Vento D, Fanfarillo A (2019) Traps, pitfalls and misconceptions of machine learning applied to Scientific Disciplines. In: Proceedings of the practice and experience in advanced research computing on rise of the machines (learning), pp 1–8
19. Hill DJ, Minsker BS (2010) Anomaly detection in streaming environmental sensor data: A data-driven modeling approach. Environmental Modelling & Software 25:1014–1022
20. Hill DJ, Minsker BS, Amir E (2007) Real-time Bayesian anomaly detection for environmental sensor data. In: Proceedings of the 32nd Congress of IAHR, International Association of Hydraulic Engineering and Research, Venice, Italy
21. Ge S, Jun L, Liu D, Peng Y (2015) Anomaly detection of condition monitoring with predicted uncertainty for aerospace applications. In: 12th IEEE international conference on electronic measurement & instruments (ICEMI), Qingdao, pp 248–253
22. Wang X, Feng H, Fan Y (2015) Fault detection and classification for complex processes using semi-supervised learning algorithm. Chemometrics Intell Lab Syst 149. Part B 15:24–32
23. Van Engelen JE, Hoos HH (2020) A survey on semi-supervised learning. Machine Learning 109:373–440
24. Shorten C, Khoshgoftaar TM (2019) A survey on image data augmentation for deep learning. J Big Data 6(1)
25. Cavalaglio Camargo Molano J, Campo F, Capelli L, Massaccesi G, Borghi D, Rubini R, Cocconcelli M (2021) A structured approach to machine learning for condition monitoring: a case study. In: Smart monitoring of rotating machinery for industry 4.0, theory and applications. Springer Applied Condition monitoring book series
26. Triguero I, Galar M, Vluymans S, Cornelis C., Bustince, H., Herrera, F., Saeys, Y., Evolutionary undersampling for imbalanced big data classification. 2015 IEEE Congress on Evolutionary Computation (CEC)
27. Li J, Cheng K, Wang S, Morstatter F, Trevino RP, Tang J, Liu H (2017) Feature selection: a data perspective. ACM Comput Surv (CSUR), 5(1–45)
28. Xu S, Lu B, Baldea M, Edgar TF, Wojsznis W, Blevins T, Nixon M (2015) Data cleaning in the process industries. Reviews in Chemical Engineering 31(5):453–490
29. Khalid S, Khalil T, Nasreen S (2014) A survey of feature selection and feature extraction techniques in machine learning. In: 2014 Science and Information Conference, pp 372–378
30. Liang H, Sun X, Sun Y, Gao Y (2017) Text feature extraction based on deep learning: a review. EURASIP journal on wireless communications and networking 1:1–12
31. Delac K, Grgic M, Grgic S (2005) Independent comparative study of PCA, ICA, and LDA on the FERET data set. International Journal of Imaging Systems and Technology 15(5):252–260
32. Patel RK, Giri VK (2016) Feature selection and classification of mechanical fault of an induction motor using random forest classifier. Perspect Sci 8:334–337
33. Lei Y, Yang B, Jiang X, Jia F, Li N, Nandi AK (2020) Applications of machine learning to machine fault diagnosis: a review and roadmap. Mech Syst Signal Process 138

34. Muralidharan K (2010) A note on transformation, standardization and normalization. Int J Oper Quant Manage IX(1-2):116–122
35. Widodo A, Bo-Suk Y (2007) Support vector machine in machine condition monitoring and fault diagnosis. Mechanical systems and signal processing 21(6):2560–2574
36. Li C, Sanchez RV, Zurita G, Cerrada M, Cabrera D, Vásquez RE (2016) Gearbox fault diagnosis based on deep random forest fusion of acoustic and vibratory signals. Mech Syst Signal Process 76:283–293
37. Babu GS, Peilin Z, Xiao-Li L (2016) Deep convolutional neural network based regression approach for estimation of remaining useful life. In: International conference on database systems for advanced applications. Springer, Cham
38. Musoro JZ, Zwinderman AH, Puhan MA, Riet G, Geskus RB (2014) Validation of prediction models based on lasso regression with multiply imputed data. BMC Med Res Methodol 14(116)
39. Understanding AUC-ROC Curve https://towardsdatascience.com/understanding-auc-roc-curve-68b2303cc9c5

A Structured Approach to Machine Learning for Condition Monitoring: A Case Study

Jacopo Cavalaglio Camargo Molano, Federico Campo, Luca Capelli, Giulia Massaccesi, Davide Borghi, Riccardo Rubini, and Marco Cocconcelli

Abstract This chapter details the application of a machine learning condition monitoring tool to an industrial case study. The process follows the content of the corresponding tutorial chapter and is a step-by-step example of the setup of a monitoring kit in a packaging machine. The case study is particularly interesting since it is focused on Independent Carts System. This consists of a closed path made up of modular linear motors having a straight or curved shape and controls a fleet of carts independently. The application, not so common nowadays, proves the feasibility of the proposed condition monitoring approach in a non-trivial case, with scanty literature on it. The target is the diagnostics of ball bearings present in the wheels of the carts in order to reduce downtime due to the breakage of these components and to maximize their life cycle cutting down spare part costs. This chapter details the phase of feature extraction, the Machine Learning methods used, the results and the metrics for measuring them. Considerations will be made in particular on the acceptability/interpretability of the results and the industrial significance of the metrics.

J. C. C. Molano · R. Rubini · M. Cocconcelli (✉)
University of Modena and Reggio Emilia, Via G. Amendola 2, 42122 Reggio Emilia, Italy
e-mail: marco.cocconcelli@unimore.it

J. C. C. Molano
e-mail: jacopo.cavalagliocamargomolano@unimore.it

R. Rubini
e-mail: riccardo.rubini@unimore.it

F. Campo · L. Capelli · G. Massaccesi · D. Borghi
Tetra Pak Packaging Solutions, Via A. Delfini 1, 41123 Modena, Italy
e-mail: federico.campo@tetrapak.com

L. Capelli
e-mail: luca.capelli@tetrapak.com

G. Massaccesi
e-mail: giulia.massaccesi@tetrapak.com

D. Borghi
e-mail: davide.borghi@tetrapak.com

F. Chaari et al. (eds.), *Smart Monitoring of Rotating Machinery for Industry 4.0*,
Applied Condition Monitoring 19, https://doi.org/10.1007/978-3-030-79519-1_4

1 Introduction

The improvement of technology, especially as regards logic controllers, has enabled linear servo motors to perform tasks that were not possible before. In this way, mechanical complexity has decreased while software complexity has increased.
The Independent Carts System (Figs. 1 and 2) uses linear motors to control one or more movers that are constrained by rollers to follow a track. The track can have different shapes with curved and straight parts and it has a flexible architecture in order to build modular configurations. In this way, a high-performance flexible system can be carried out, with this technology each mover can be controlled independently [1, 2].The movers can accelerate, decelerate, take an absolute position and produce forces. The velocity of the movers can be very high with respect to rotary motors, each mover can move with a velocity of 4m/s.

Thanks to the reduction of moving parts, the maintenance of the components is reduced with respect to the technology of chain and belt drives. The bearings inside the rollers are subject to wear and the condition monitoring of this system is challenging due to the non-stationary working conditions. As a matter of fact each cart can have a different motion and load profile that can also change during the production. The objective of the chapter is to show the use of different Machine Learning algorithms to detect the damaged bearings. Technical backgrounds and details about machine learning and its application for condition monitoring can be find in the

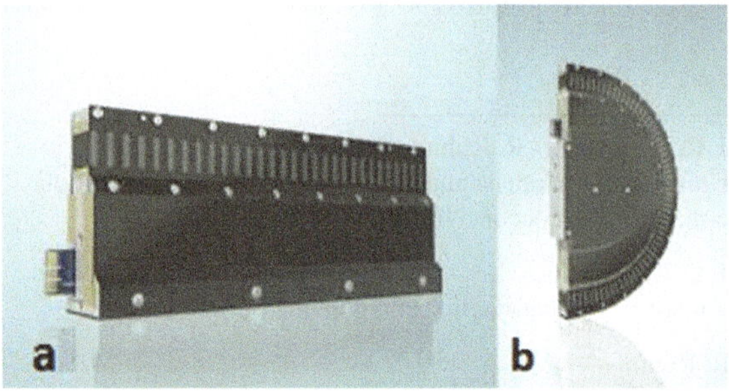

Fig. 1 Straigth and curved linear motors

Fig. 2 XTS system with 12 carts

chapter 3 titled A Structured Approach to Machine Learning for Condition Monitoring, in this same book [3]. The system taken into account is an XTS Beckhoff System and the use of Machine Learning algorithms allows to predict the time when a single cart has to be replaced. This work is organized as follows: Sects. 2 and 3 describe the Data driven algorithms used for the implementation of the fault detection classifiers. Section 4 describes the different experiments, the training and the validation of the different models. Section 5 explains the conclusions of the research.

2 Random Forest

The Random Forest (RF) algorithm was developed by Breiman [4] in 2001. Random Forest is a machine learning model based on tree bagging. Bagging is a machine learning ensemble meta-algorithm designed to improve stability and accuracy; it is used in statistical classification and regression. It also reduces variance and helps to avoid overfitting. Random Forest consists of an ensemble of simple decision-tree predictors, each of which gives a class prediction as output and the class that has the largest number of votes becomes the prediction of the model (Fig. 3).

In the nodes of the trees there are thresholds based on one or more features that decide if the data must proceed to the left or to the right of the tree. On the contrary, as regards the leaves, the probabilities are calculated on the basis of the elements of each class that ends up in a given leaf. As regards classification problems, the ensemble of simple trees votes for the most popular class. As regards regression problems, the responses of the trees are averaged to obtain an estimate of the depen-

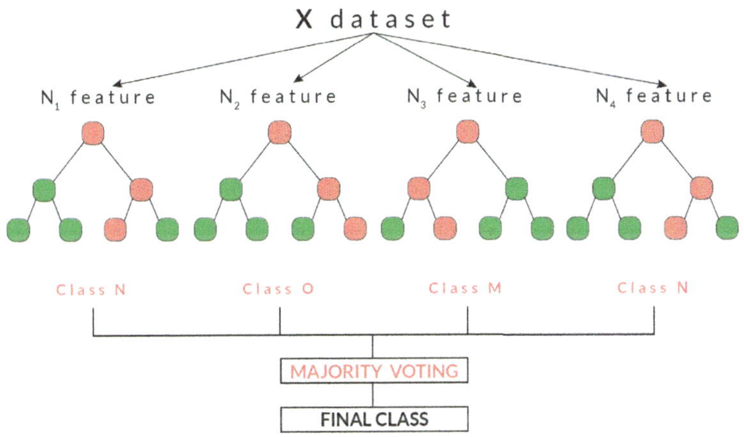

Fig. 3 Random forest Classifier architecture [7]

dent variables. The use of tree ensembles can lead to a significant improvement in prediction accuracy.

3 Deep Learning/Autoencoder

Autoencoders were first introduced in the 80s by Hinton and the PDP group [5] to solve the problem of "backpropagation without a teacher". The aim was to train a model on a dataset with no pre-existing labels and with the minimum human supervision. Autoencoders are neural networks, used with the purpose of generating new data, firstly by compressing the inputs into a space of latent variables present in the middle layer of the model, which are not directly observed but are rather inferred; secondly by reconstructing the output based on the acquired information. These latent variables consist in the most salient features extracted from the inputs. Autoencoders generally use neuron dimensional reduction algorithms in order to force the model to learn how to represent, with a smaller number of dimensions, the space represented by all the training set. The Autoencoder network consists of two parts:

- **Encoder**: the part of the network that compresses the inputs into a space of latent variables and which can be represented by the encoding function:

$$h = f(x) \tag{1}$$

- **Decoder**: the part that deals with the reconstructing of the inputs on the basis of the information previously collected. It is represented by the decoding function:

$$r = g(h) \tag{2}$$

Therefore, the Autoencoder, taken as a whole, can be described by the function

$$d(f(x)) = r \tag{3}$$

where r is the most similar to the original input x.

There are several variants on the basic model, which aim to make the learned representations of the inputs assume useful properties. In this application, the Autoencoder is mainly used for anomaly detection, by learning to replicate the most salient features in the training data. The model is encouraged to learn how to reproduce the most frequent characteristics of the observations precisely. When facing anomalies, the model worsens its reconstruction performance. In most cases, only data with normal instances are used to train the Autoencoder (Fig. 4).

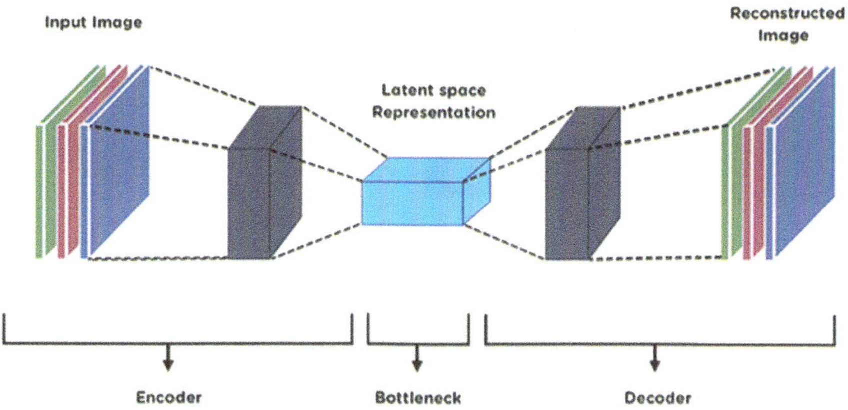

Fig. 4 Autencoder structure [6]

4 Problem Description

Because of the newness of the Independent Carts Systems and the scarcity of these machines in a real production plant, there is not yet any knowledge of real damages of the bearings for the XTS system. Consequently, fictitious damages were created in order to develop a data-driven monitoring system. They are as similar as possible to real damages that can occur in the field.

The damages created are the following:

- **Rusty damage**: it is created by immersing the bearing into a solution of water and salt for one week. It is considered as a distributed damage.
- **Inner race damage**: it is created by drilling the inner ring with a tip of 0:2 mm. It is the lightest damage and it is punctual.
- **Outer race damage**: it is created by cutting the outer ring of the bearing. It is a serious damage and it is punctual.
- **Blockage damage**: it is created by blocking the sphere of the bearing with rust and metals. It is the most serious damage; it is also very dangerous for the rail, if it is not recognized quickly (Fig. 5).

After the creation of these four types of damages, it was observed that the rusty bearing could not rotate. For this reason, rust was removed and the bearing was lubricated in order to give it the possibility of spinning. In the image below, it is possible to see the transformation (Fig. 6).

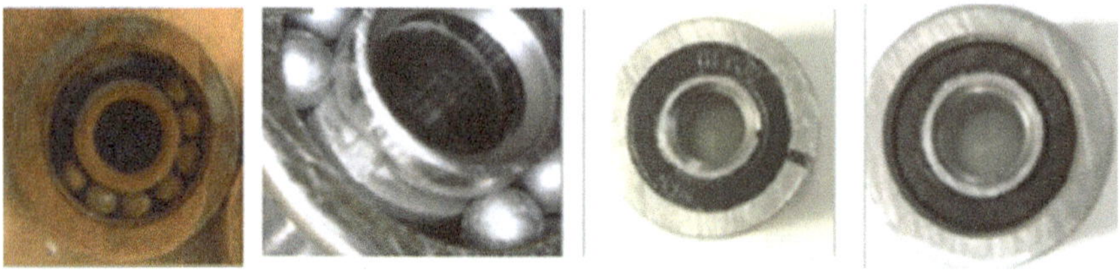

Fig. 5 Damaged bearing images, from the left to the right: (1) Rusty bearing, (2) Bearing with an inner ring damage, (3) Bearing with an outer ring damage, (4) Blocked bearing

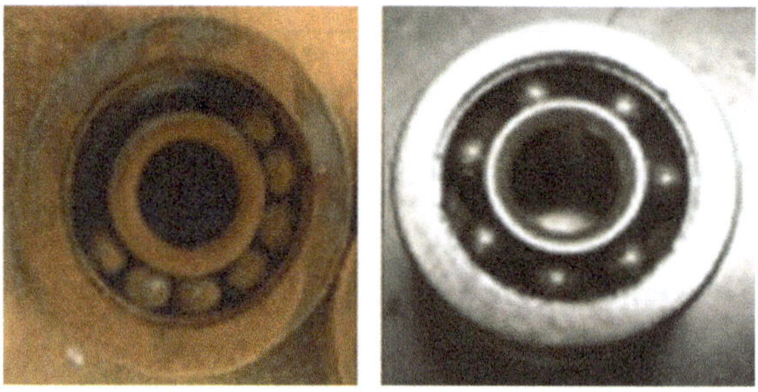

Fig. 6 From the left to the right: (1) Bearing during water treatment (2) Rusty bearing after the removal of the rust

4.1 Preliminary Test on Rotary Test Rig

A simplified and hyper-monitored test rig was used to guarantee the correctness of the artificial damages. The term correctness means the possibility of observing the damages with machine learning algorithms in a simpler case. This first test also helps to check if the artificially damaged bearings are too dangerous for the entire Independent Carts System. In this case, the simplified tests consist in the analysis of the inner, outer, rusty, blocked damaged bearings and of the healthy one in the rotary motor test rig (Fig. 7)

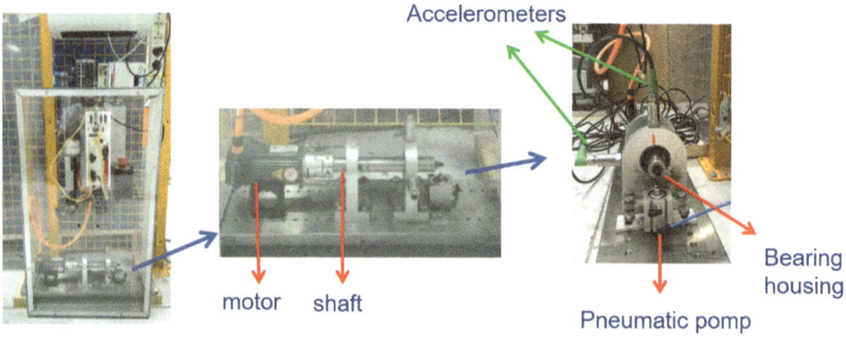

Fig. 7 Rotary motor test rig

The Rotary Motor Test Rig is made up of:

- Servo motor Beckhoff AM8022-0D20-0000
- Shaft connected to the motor on which the bearings are mounted.
- Piezoelectric accelerometers (IFM VSP001) placed at 90 degrees to monitor the bearing with high precision.
- PLC and driver for the control of the motor and the recording of the accelerometer signals and the motor signals.

The variables recorded for the tests are the following:

1. The vibrations of the lateral and upper accelerometers recorded for twenty-five times, with a sampling frequency of 20 kHz, each recording for one hundred seconds.
2. The current of the servomotors recorded for twenty-five times, with a sampling frequency of 400 Hz,each recording for one hundred seconds.

Considering that in this test the bearing velocity is equal to 300RPM, about 2500 samples were generated for each type of condition (rusty, inner race damage, outer race damage, blockage damage and healthy). This dataset has been divided by using 1500 samples of each condition in the training set and 1000 samples of each condition in the test set. The raw vibration and current data were recorded for the different typologies of the bearings installed on the Rotary Motor Test Rig and a labelled dataset was developed. Several features in time domain and frequency domain were computed for each recording, they were used for the development of a Random Forest Classifier. The computed features are listed in Appendix. A Random Forest Classifier was trained and tested to observe if the differences between the damaged and healthy bearings were detectable. Figure 8 shows the weight that the Random Forest Classifier gives to the first ten most relevant variables. The higher the weight, the more important the feature for the classification in a precise configuration of the Random Forest. To have a complete map of the importance of the features and to find the most significant ones, several training datasets with different subsets of features were given to the Random Forest.

The results of these Random Forest models can be summarized by using the confusion matrix and accuracy, precision and recall (Fig. 9). Backgrounds and details about reconstruction error and confusion matrix can be find in [3] (Fig. 10).

From these results, it is possible to deduce that the artificially damaged bearings are detectable by means of the Random Forest model trained with the vibration variables or with the current variables. Figure 11 shows the weight that the Random forest Classifier gives to the first ten most relevant variables but considering current signals only. The tests show that the current signal gives less information about the condition of the bearings, most probably due to the sampling rate and the control loop of the driver. De facto, the current sampling rate is 400 Hz, a lower value in comparison with the sampling rate of vibrations that is equal to 20 kHz. Consequently, the current signal does not allow to analyze if there are resonance phenomena at frequencies higher than 200 Hz, while the vibration signals allow to analyze resonance

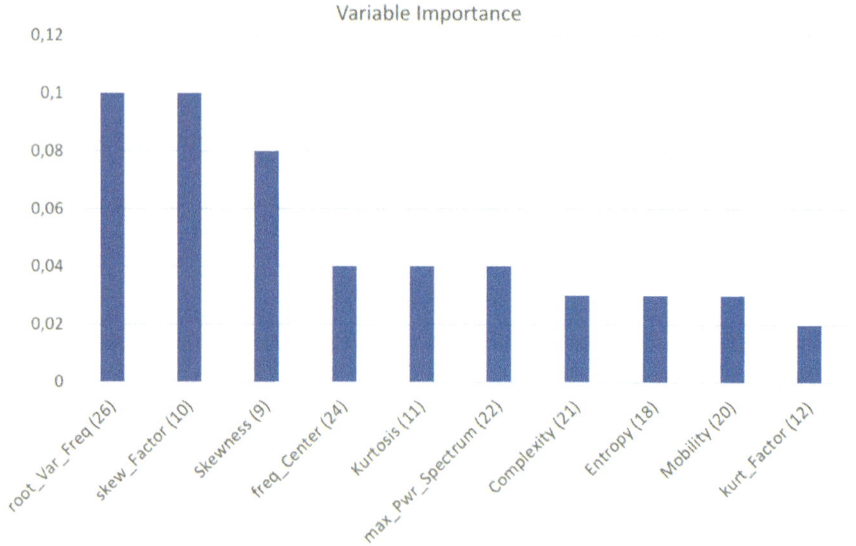

Fig. 8 The importance of all the features evaluated by the Random Forest algorithm. Numbers in brackets refer to specific parameter detailed in Appendix

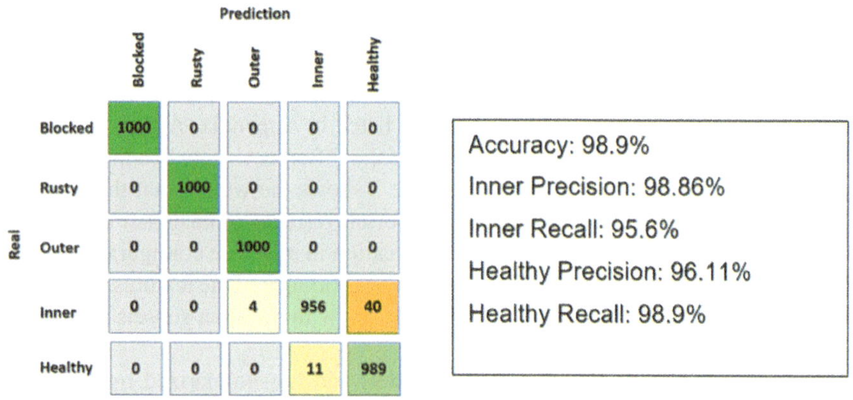

Fig. 9 Confusion matrix of the random forest algorithm in the test set, trained on all the features

phenomena up to 10 kHz. The other reason is due to the motor control loop, because the internal P.I.D of the driver directly changes the supply current of the motor and this action is overlapped to the current variation due to the bearing damage. Though there are these drawbacks, the current signal allows to identify the condition of the bearings, even if with less accuracy.

This separate analysis of current and vibration was made because the signals are recorded in both Rotary Test Rig and XTS Test Rig. Because of the differences between the two test rigs, the significance of the two signals changes chiefly because

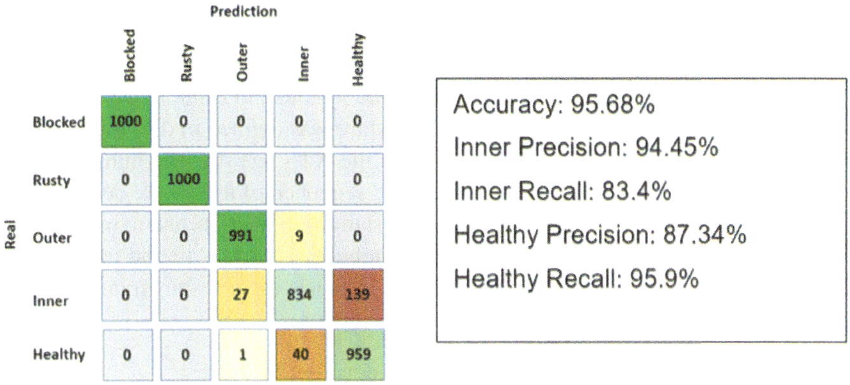

Fig. 10 Confusion Matrix of the Random Forest algorithm in the test set, trained only on the current features

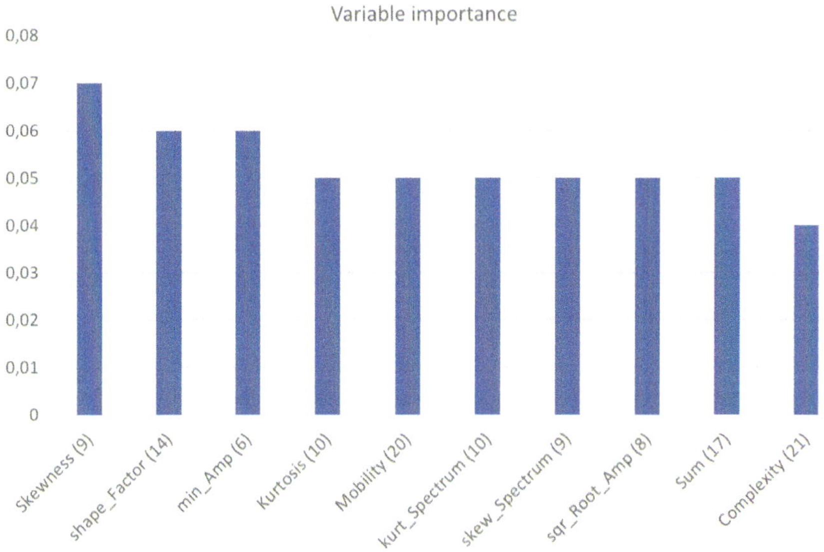

Fig. 11 Importance evaluated by the Random Forest algorithm trained only with current features. Numbers in brackets refer to specific parameter detailed in Appendix

in the Rotary Test Rig the accelerometers are placed close to the bearings, while in the XTS Test Rig they are placed on the fixed frame of motors. De facto, in the XTS Test Rig the vibrations are not directly linked to each cart and consequently they reduce their significance. Furthermore, it is possible to observe that, regardless of the pre-processed signals, all the trained Random Forest models give more importance to Skewness, RMS, Kurtosis in comparison with the other features. These three features are already used in the literature on condition monitoring and this is a further proof of the correct implementation of the Random Forest models.

4.2 XTS Test Rig

With the certainty that the artificial damages are visible and cannot cause damages on the Independent Carts System, a test procedure was defined to develop a dataset with the aforementioned damaged bearings in the XTS test rig. The number of the movers used to test the faulty bearings and the sequence of tests were decided randomly to reduce the possible environmental variations in the tests and to improve the repeatability of the tests themselves. The total number of tests is 12 and for each test the system variables were recorded four times. The test table is shown in Fig. 12.

Each test was run following this standard procedure:

1. Setting of the faulty bearing on the mover indicated by the test.
2. Twenty-minute warming-up of the test rig without any data recording.
3. Forty-second recording of all variables considered.
4. Repetition of the procedure from point three to point four for six times.

The variables taken into consideration are:

- Actual position of each cart, with a sampling rate of 4 kHz.
- Actual velocity of each cart, with a sampling rate of 4 kHz.
- Position and velocity errors expressed as the difference between the actual position and velocity with respect the real position and velocity, all with a sampling rate of 4 kHz.
- Actual current of each cart, with a sampling frequency of 4 kHz.
- Vibration signals of two accelerometers placed on the top and bottom part of the frame with a sampling frequency of 20 kHz.

The procedure for creating the test dataset, which is shown in Fig. 12, is the same procedure used for the training dataset reported in Fig.13

It can be observed that, during the training phase, no data about the blocked bearing were collected, since this is an extremely invasive damage that can seriously damage the track of the XTS. In order to train and validate the algorithm, the data

	Number of movers											
Test number	M1	M2	M3	M4	M5	M6	M7	M8	M9	M10	M11	M12
T1	H	H	H	H	H	H	H	H	H	H	H	R
T2	H	H	H	H	H	H	H	O	H	H	H	H
T3	H	H	H	H	H	H	H	H	H	I	H	H
T4	H	H	H	H	H	R	H	H	H	H	H	H
T5	H	H	H	H	H	O	H	H	H	H	H	H
T6	H	H	H	I	H	H	H	H	H	H	H	H
T7	H	R	H	H	H	H	H	H	H	H	H	H
T8	H	O	H	H	H	H	H	H	H	H	H	H
T9	H	H	I	H	H	H	H	H	H	H	H	H
T10	H	H	H	H	R	H	H	H	H	H	H	H
T11	O	H	H	H	H	H	H	H	H	H	H	H
T12	H	H	H	H	H	H	H	H	H	H	R	H

H = Healthy State I = Inner Damage O = Outer Damage R = Rusty Damage

Fig. 12 Training table with the different tests for all the types of damages

Test number	Number of movers											
	M1	M2	M3	M4	M5	M6	M7	M8	M9	M10	M11	M12
T1	H	H	H	H	H	H	H	H	H	H	H	I
T2	H	H	H	H	R	H	H	H	H	H	H	H
T3	H	B	H	H	H	H	H	H	H	H	H	H
T4	H	H	H	H	H	H	H	H	H	O	H	H
T5	H	H	H	H	I	H	H	H	H	H	H	H
T6	H	H	R	H	H	H	H	H	H	H	H	H
T7	H	H	H	H	B	H	H	H	H	H	H	H
T8	H	O	H	H	H	H	H	H	H	H	H	H
T9	H	H	H	H	H	H	H	H	H	H	H	H
T10	H	I	H	H	H	H	H	H	H	H	H	H
T11	H	H	H	H	H	H	H	H	H	H	H	R
T12	H	H	H	H	H	H	H	H	H	H	H	B
T13	H	H	H	H	H	H	O	H	H	H	H	H
T14	I	H	H	H	H	H	H	H	H	H	H	H
T15	H	H	R	H	H	H	H	H	H	H	H	H
T16	H	H	H	H	B	H	H	H	H	H	H	H
T17	H	H	H	H	H	H	H	H	H	O	H	H
T18	H	H	H	H	H	H	H	H	H	H	H	H

H = Healthy State   I = Inner Damage   O = Outer Damage   R = Rusty Damage   B = Blocked Damage

Fig. 13 Test table with the different tests for all the types of damages

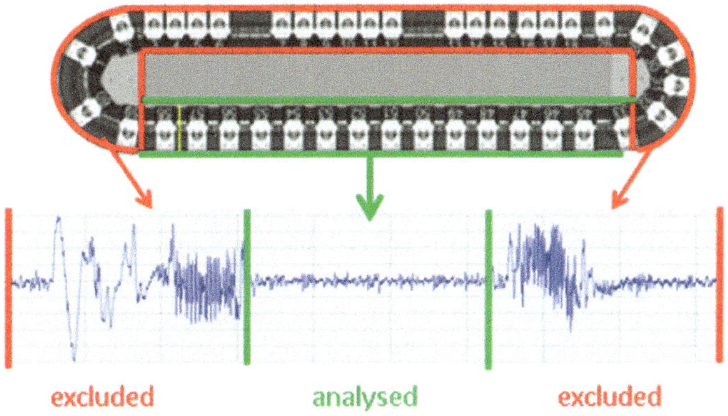

Fig. 14 Signal cut in the constant velocity part with respect to the mover position

coming from the XTS Test Rig were pre-processed. The training dataset consists in 12 different tests (Fig. 12), each of them includes 6 records of the system variables for each mover, while the test dataset consists in 18 different tests (Fig. 13), each of them includes 6 records of the system variables for each mover. For each recording, the row signals were divided into laps considering the actual position of each mover and eliminating the signals recorded along the curved parts of the track. The signals recorded in the curved parts are eliminated because they show a very high level of noise. Even the signals recorded on the top strength part of the track are eliminated because in that zone the movers have a variable motion profile that increases the complexity of the analysis. Figure 14 shows the part of the recorded signals taken into consideration after the pre-processing.

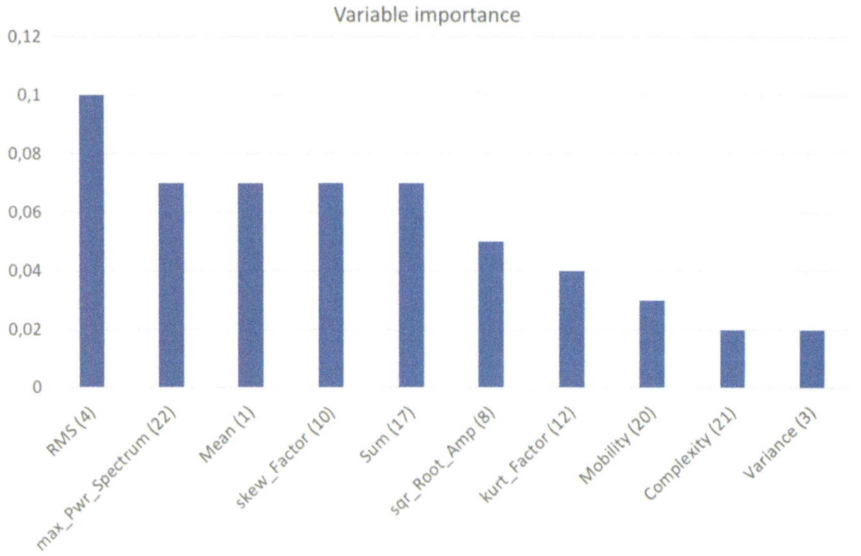

Fig. 15 Random forest weight for each feature. Numbers in brackets refer to specific parameter detailed in Appendix

The vibrations, which are not referable to the actual position of the mover since the sensors are placed on the frame and not on each cart, are cut into different laps considering position 0 of mover 1 as starting point and ending point. Carrying out different tests on these data, it is noticed that the features of vibrations are not considered by the Random Forest Classifier because they are not referred to each mover but to the general system. It can also be noted that the algorithm, trained with signals recorded in the upper part of the track, tend to overfit the prediction on each mover. This can be explained considering that the carts have variable velocities in the upper part of the track and the motion profile is different cart by cart, while they have a constant velocity in the bottom part. To avoid the aforementioned problems, the vibration data are not taken into account, but only the other signals recorded in the bottom zone of the test rig are considered. For each pre-processed signal, the features of Appendix are calculated.It is important to try to understand the reasons of the importance given to the features by the Random Forest. This allows to carry out a robust pre-processing phase, in which it is necessary to select the features useful to solve the problem, without considering the ones that are only descriptive of the training set and do not generalize the problem. This leads to avoid the overfitting of the model.

As regards the scarcely meaningful features, the algorithm gives them a very low weight, so they are directly discarded. Fig. 15 shows the importance of the first ten most relevant features evaluated by the Random Forest. It is possible to notice that the features referred to vibrations are not considered, while the ones referred to current are the most significant for prediction.

Test number	Number of movers											
	M1	M2	M3	M4	M5	M6	M7	M8	M9	M10	M11	M12
T1	H	H	H	H	H	H	H	H	H	H	H	I
T2	H	B	H	H	H	H	H	H	H	H	H	H
T3	H	H	H	H	H	H	H	H	H	O	H	H
T4	H	H	H	H	I	H	H	H	H	H	H	H
T5	H	H	H	H	B	H	H	H	H	H	H	H
T6	H	O	H	H	H	H	H	H	H	H	H	H
T7	H	H	H	H	H	H	H	H	H	H	H	H
T8	H	I	H	H	H	H	H	H	H	H	H	H
T9	H	H	H	H	H	H	H	H	H	H	H	B
T10	H	H	H	H	H	H	O	H	H	H	H	H
T11		H	H	H	H	H	H	H	H	H	H	H
T12	H	H	H	H	B	H	H	H	H	H	H	H
T13	H	H	H	H	H	H	H	H	H	O	H	H
T14	H	H	H	H	H	H	H	H	H	H	H	H

Green = true healthy
Yellow = false faulty
Red = true faulty
Violet = false healthy

Inner detected = 2/4
Outer detected = 4/4
Blocked detected = 4/4

Total of fault detected= 10/12

Fig. 16 Binary Random Forest Classifier final output

It was also evaluated the possibility to use only the ten most important features for the Random Forest training and test. The comparison between the Random Forest Classifier based on the ten most important features and the Random Forest Classifier that uses all the features, except vibrations, shows a higher accuracy prediction in the second model. For this reason, the output data shown in Fig. 17 regard the Random Forest Classifier trained with all the features, except vibrations. In order to evaluate the correctness of the trained algorithm for an industrial target, it is not possible to take into account only accuracy, recall and precision because it is necessary also to evaluate the cost of false predictions (false positives) and missed alarms (false negatives). De facto, if the system under monitoring is crucial and the cost of sending a service engineer to monitor it, in case of alarm, is low, it is better to have an algorithm with the lowest number of false negatives, but it is acceptable to have a large number of false positives. Nevertheless, the opposite case can occur as regards both the terms of cost and the conditions of the system. For this reason, a good approach is to evaluate the cost of the possible errors of the algorithm and give them a weight dependent on the cost. There is no information on the costs and consequently it was chosen to create a variable in order to tune the model according to the needs. The "tuned variable" consists in defining a threshold of the number of predictions referring to faulty bearings before the system predicts a damage by means of an alarm. The threshold was established on the following basic rule:

- if the model classifies as faulty more than 35% of the recordings regarding a mover in a given test, the global prediction is considered faulty.

Figure 16 shows the results of the binary classification with the Random Forest Classifier and Fig. 17 shows the complete confusion matrix. In this case the rusty bearings were not taken into consideration because they did not show a level of damage that could reduce the performances of the machine

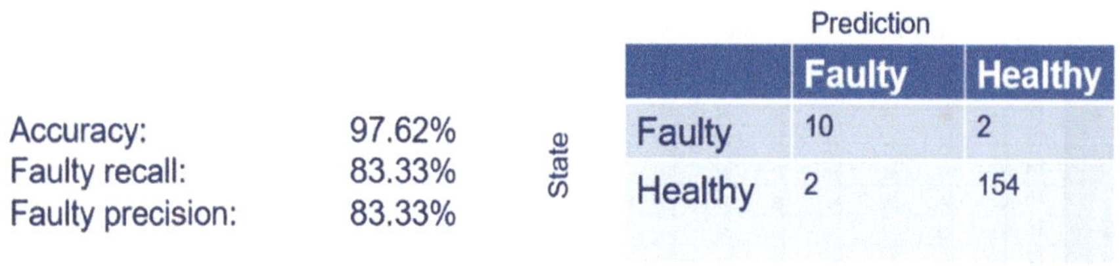

Fig. 17 Binary Random Forest Classifier Confusion Matrix

4.3 Autoencoder for Anomaly Detection

In order to have an outline of different data-driven techniques, it was chosen to use an Autoencoder model to monitor the state of the mover bearings of the XTS system. In this case, the Autoencoder was used for reconstructing the healthy state of the carts in the best way. De facto, when a bearing suffers a damage, the reconstruction error increases and by means of this output it is possible to detect the problem. For this reason, the training datasets of the Autoencoder include only signals of the healthy carts, while the test datasets include the signals of both healthy and damaged carts (Fig. 18).

In this case, the data used for training and test are the raw data of current, divided by cart loops on the XTS path, considering only the bottom part of the path as shown in Fig. 14. The new training set consists in healthy data, while the test set consists in all the faulty data plus some healthy data that are not used in the training dataset. In deep learning models the following functions are greatly important:

- **Loss Function**: it is the function that has to be minimized or maximized by the algorithm, it is also called cost function or error function.
- **Evaluation Function**: it is the function used to compare the real input with it is own reconstruction, during test phase.

In this specific case, for the training and test of the Convolutional Autoencoder 1D, the Mean Squared Error was used both as Loss Function and Evaluation Function. The main characteristics of the model are listed and briefly explained below:

- Convolutional Autoencoder 1D: in this case the samples are raw data with one dimension, for this reason this type of Autoencoder was chosen. It is not the

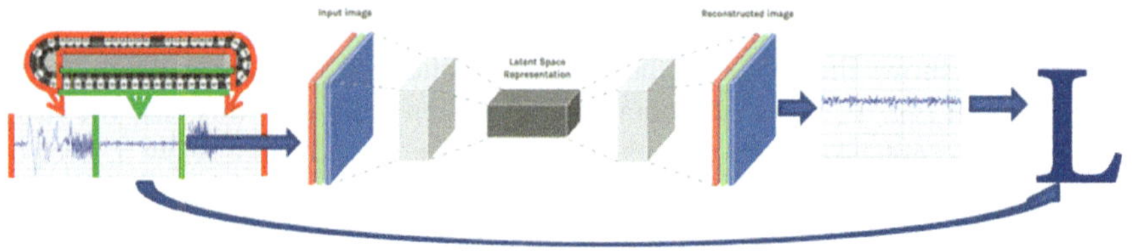

Fig. 18 Convolutional Autoencoder 1D as Anomaly detector, for faulty carts discovery

most common case of Convolutional Neural Network (CNN) utilization, since it is usually used for Images (CNN 2D).

- Mean Squared Error or Reconstruction Error: it was used to make the model converge to a local minimum during training. The more the model reduces it during the descent phase of stochastic gradient and the more the input is similar to the output. The more the Loss value of a model is low and the more the features in the latent space (middle layer of Autoencoder) describe the input correctly.

$$MSE = \frac{1}{N}\sum_{i=1}^{N}(f_i - y_i)^2 \tag{4}$$

- Padding of type Causal: It allows not to violate the temporal order of the samples with respect to the normal padding that violates this constraint. In this case it has been very useful, because the samples utilized in training and test of the model are temporal data.

Figures 19, 20 and 21 shows the reconstruction error distributions for the different types of bearing conditions in the test set. A smaller reconstruction error is expected for the healthy samples, while a higher one is expected for the different types of damages. That is why the Autoencoder used was trained to recognize only the salient features of the healthy state. It can be noted that the probability distribution of the reconstruction errors for the same type of damage and cart is the Gaussian one. The Gaussian distribution of the reconstruction errors of the healthy samples is the one with the lowest mean. The distribution of reconstruction errors of rusty samples overlaps the healthy one. This is due to the fact that, after the cleaning, the rusty bearing was very closed to the healthy state. It is instead observable that the distributions of reconstruction errors of the inner, outer and blocked damages do not completely overlap the healthy one. In this way, the Autoencoder allows to generate thresholds able to separate the healthy samples from the faulty ones.

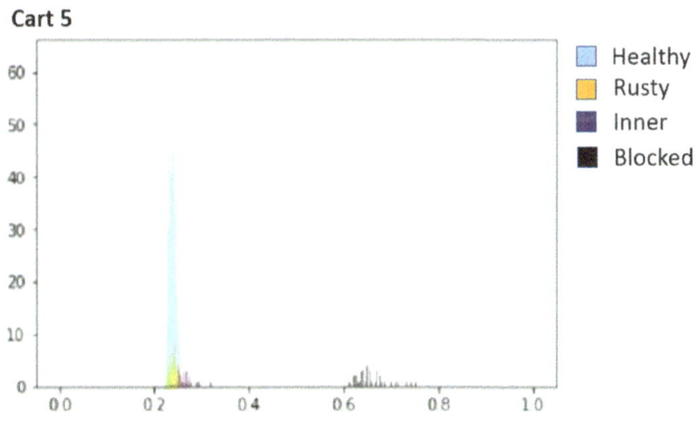

Fig. 19 Autoencoder reconstruction errors of the mover 5

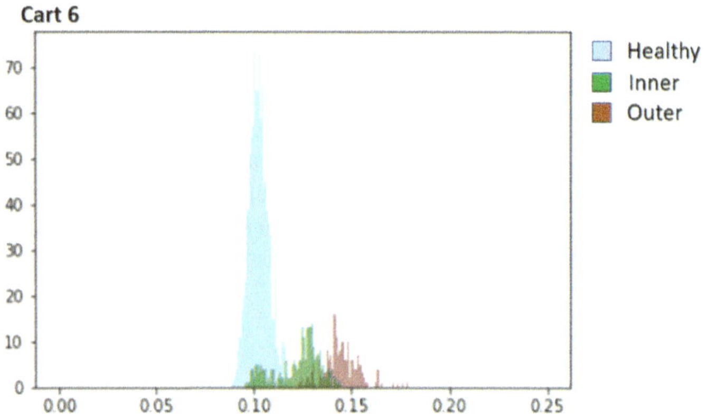

Fig. 20 Autoencoder reconstruction errors of the mover 5

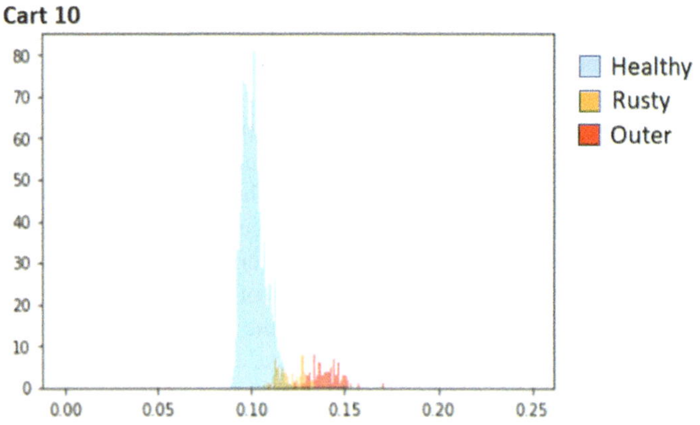

Fig. 21 Autoencoder reconstruction errors of the mover 5

Figure 22 shows the results of regression expressed in Confusion Matrix, while Fig 23 shows the percentages of the samples correctly classified by using a threshold equal to 0.13 as a discriminator between healthy and faulty samples on the top of the probabilities calculated by the Autoencoder.

As noticed in Figs. 19, 20 and 21, not all the probability distributions of movers with healthy bearings are centred on the same value. This is the factor that introduces the misclassification of some healthy samples as faulty, while the misclassification of faulty bearings as healthy is due to the small entity of the damages, which makes faulty bearings similar to the healthy ones.

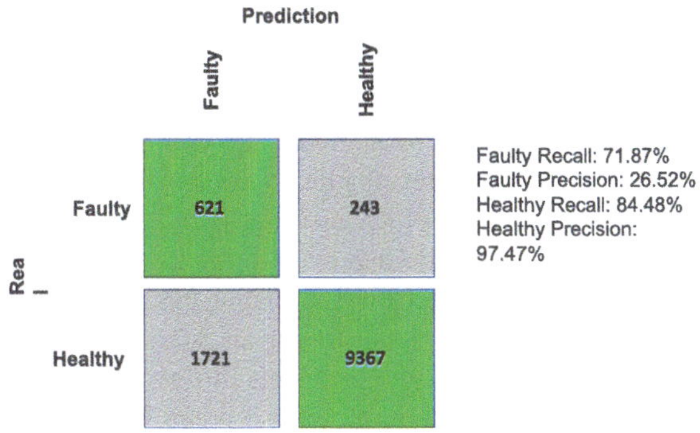

Fig. 22 Confusion Matrix of the Anomaly Detector without rusty samples

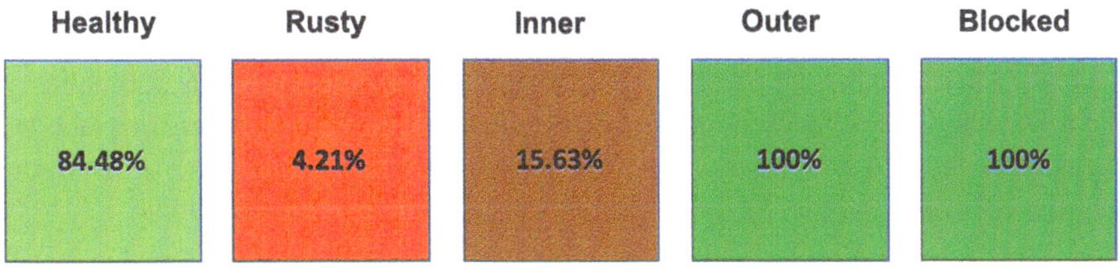

Fig. 23 Percentages of samples correctly classified by using a threshold equal to 0.13 as a discriminator between healthy and faulty samples on the top of the probabilities calculated by the Autoencoder

5 Conclusions

Since the system taken into account in this specific case is technologically recent, there is not yet any case of real damaged bearings. In order to overcome the problem, a methodology, which uses artificial damages, has been created. It allows to explore data-driven solutions even in the early stages of the development of condition monitoring infrastructures. Both Random Forest and CNN Autoencoder can classify the different types of damages and the healthy state of bearings correctly, with good precision, recall and accuracy. De facto, the trained model is considered robust. It is not too sensitive to the system variations, and identifies bearings, which are slightly worn, as damaged. In the case under study, for example, the inner damage of bearings is very light, and it does not affect the system performances. For this reason, the trained model is very accurate to detect outer ring damaged bearings and blocked bearings, but it is less accurate to detect inner damaged bearings. The artificial damages were created to be the most similar as possible to the real ones, there is no certainty that the machine bearings will have the same type of wear. Consequently, three different types of failure bearings with different levels of damages were created. Furthermore, artificial damages do not show the wear evolution of the component over time, but they represent only a definite state of damage. On the contrary, by monitoring the system in the field, it is possible to observe the evolution of real damages. This fact reduces the reliability of the above described models for industrial applications. Nevertheless, it was essential to generate them in order to understand the most relevant system variables, the way of cleaning and pre-processing the signals and to find the most promising algorithms for the detection of the system failures. As a matter of fact, the trained Random Forest model allows to identify both the harsh artificial damages, such as block and outer ring damages, and the lightest artificial damages, such as the inner ring damages, with a good accuracy. On the basis of these performance data, it is possible to conclude that even in the case of real damages there are considerable chances of training a robust classifier based on the architecture previously shown. Moreover, it is convenient to use an Autoencoder model as an Anomaly Detector in order to overcome the problems observed in the Random Forest Classifier. This type of model was chosen because it can be trained only with the class of healthy cases, encouraging it to replicate the most significant features of this class. In this way, when the machine is in the field, it will be possible to train an Autoencoder during the first months of operation and subsequently to detect possible anomalies of the cart behaviour. When the number of machines is relevant and there are labelled data of healthy and faulty carts, it will be possible to train a Random Forest Model in order to detect and identify the different types of the machine failures.

Appendix

Let S be a signal composed of K points of amplitude x_i:

1. **Mean**: it is the average of all values of the signal/sample

$$x_m = \frac{1}{K} * \sum_{i=1}^{K} x(i) \tag{5}$$

2. **Standard Deviation**: it is the deviation from the mean of the signal/sample.

$$x_{std} = \sqrt{\frac{\sum_{i=1}^{K} (x(i) - x_m)^2}{K - 1}} \tag{6}$$

3. **Variance**: it is the square of Standard Deviation.

$$x_{var} = \frac{\sum_{i=1}^{K} (x(i) - x_m)^2}{K - 1} \tag{7}$$

4. **Root Mean Square**: it is the square root of the mean of squares of a signal/sample.

$$x_{rms} = \sqrt{\sum_{i=1}^{K} \frac{x(i)^2}{K - 1}} \tag{8}$$

5. **Maximum Amplitude**: it is the value of the maximum amplitude of the signal/sample.

$$x_{max} = max(x(i)) \tag{9}$$

6. **Minimum Amplitude**: it is the value of the minimum amplitude of the signal/sample.

$$x_{min} = min(x(i)) \tag{10}$$

7. **Peak to Peak Value**: it is the difference between maximum and minimum peak values.

$$x_{ppv} = x_{max} - x_{min} \tag{11}$$

8. **Square Root of Amplitude**: it is the value of the root of Amplitude.

$$x_{sra} = (\frac{1}{K}\sum_{i=1}^{K}\sqrt{|x(i)|})^2 \tag{12}$$

9. **Skewness**: it is the measure of lack of symmetry in the probability distribution function.

$$x_{skew} = \frac{\sum_{i=1}^{K}(x(i) - x_m)^3}{(K-1)x_{std}^3} \tag{13}$$

10. **Skewness Factor**: it is the Skewness value divided by the square of the mean of squares of amplitudes.

$$x_{skewFactor} = \frac{\frac{\sum_{i=1}^{K}(x(i)-x_m)^3}{(K-1)x_{std}^3}}{(1/K\sum_{i=1}Kx(i))^3} \tag{14}$$

11. **Kurtosis**: it is the measure of the spikiness of the signal/sample relative to a normal distribution.

$$x_{kurt} = \frac{\sum_{i=1}^{K}(x(i) - x_m)^4}{(K-1)x_{std}^4} \tag{15}$$

12. **Kurtosis Factor**: it is the Kurtosis value divided by the square of the mean of squares of amplitudes.

$$x_{kurtFactor} = \frac{\frac{\sum_{i=1}^{K}(x(i)-x_m)^4}{(K-1)x_{std}^4}}{(1/K\sum_{i=1}Kx(i)^2)^2} \tag{16}$$

13. **Clearance Factor**: it is the ratio of maximum amplitude value to square of mean of root of absolute values.

$$x_{clf} = \frac{x_{max}}{(1/K \sum_{i=1}^{K} \sqrt{|x(i)|})^2} \tag{17}$$

14. **Shape Factor**: it is the value of how much the shape of a signal is affected, other than shifting or scaling.

$$x_{sf} = \frac{x_{rms}}{(1/K \sum_{i=1}^{K} \sqrt{|x(i)|})} \tag{18}$$

15. **Impulse Factor**: it is the ratio of maximum amplitude value to mean of absolute values.

$$x_{if} = \frac{x_{max}}{1/K \sum_{i=1}^{K} |x(i)|)} \tag{19}$$

16. **Crest Factor**: it is the ratio between the maximum amplitude and the RMS value of the signal/sample.

$$x_{cf} = \frac{x_{max}}{xmin} \tag{20}$$

17. **Sum**: it is the sum of all signal point values in a sample/signal.

$$x_{sum} = \sum_{i=1}^{K} x(i) \tag{21}$$

18. **Entropy**: it is a calculation of the uncertainty and randomness of a sampled signal.Given a set of probabilities, $(p1, p2, ..., pn)$, the entropy can be calculated as:

$$e(p) = -\sum_{i=1}^{K} p(z_i)log2p(z_i) \tag{22}$$

19. **Activity**: it is the variance of the signal.

$$Activity = \sigma_x^2 \tag{23}$$

20. **Mobility**: it is the square root of the ratio of the activity of the first derivative and the activity of the vibration signal.

$$Mobility = \frac{\sigma_x^{'}}{\sigma_x} \tag{24}$$

where $\sigma_x^{'}$ is the standard deviation of the first derivative of the vibration signal.

21. **Complexity**: it is calculated as the ratio of mobility of the first derivative and the mobility of the vibration signal.

$$Complexity = \frac{\frac{\sigma_x^{''}}{\sigma_x^{'}}}{\frac{\sigma_x^{'}}{\sigma_x}} \quad (25)$$

22. **Max Power Spectrum**: it is the Value of the maximum power of the frequency spectrum.

$$x_{fmax} = max(Power(n)) \quad (26)$$

23. **Max Envelope**: it is the maximum value of the envelope of the signal/sample.

$$x_{env} = max(Env) \quad (27)$$

24. **Frequency Center**: it is the average of all values of spectrum of the signal/sample.

$$f_c = \frac{\sum_{i=1}^{K} f * S(n)}{\sum_{i=1}^{K} S(n)} \quad (28)$$

25. **Root Mean Square Frequency**: it is the square root of the mean of squares of spectrum of a signal/sample.

$$f_{rms} = \frac{\sqrt{\sum_{i=1}^{K} f^2 * S(n)}}{\sum_{i=1} KS(n)} \quad (29)$$

26. **Root Variance Frequency**: it is the deviation from the center of the frequency of the signal/sample.

$$f_{std} = \frac{\sqrt{\sum_{i=1}^{K} (f - f_c)^2 * S(n)}}{\sum_{i=1}^{K} S(n)} \quad (30)$$

References

1. Beckhoff Automation. XTS. The eXtended Transport System https://download.beckhoff.com/download/Document/Catalog/XTS_Beckhoff_e.pdf
2. B&R: ACOPOStrak Ultimate Production Effectiveness. https://tinyurl.com/yd6myql2 https://www.br-automation.com/smc/5adafdb3a7f954f17c8bb25652a8c971a38e4d94.pdf

3. Capelli L, Massaccesi G, Cavalaglio Camargo Molano J, Campo F, Borghi D, Rubini R, Cocconcelli M (2021) A structured approach to machine learning for condition monitoring. In: Smart monitoring of rotating machinery for industry 4.0, theory and applications. Springer Applied Condition monitoring book series
4. Breiman L (2001) Random forest. Mach Learn 45:5–32
5. Rumelhart DE, Hinton GE, Williams RJ (1986) Learning representations by back-propagating errors. Nature 6088:533–536
6. https://hackernoon.com/. Cited 16 Jul 2020
7. https://blog.quantinsti.com/random-forest-algorithm-in-python/

Dynamic Reliability Assessment of Structures and Machines Using the Probability Density Evolution Method

Sajad Saraygord Afshari, Ming J. Zuo, and Xihui Liang

Abstract The reliability of a structure or machine is affected by many factors, such as operating conditions and design quality. Dynamic reliability assessment of structures and machines has been an important research topic, and many researchers have tried to address this problem. The Probability Density Evolution Method (PDEM) is a practical approach for accurate reliability assessment, especially when dealing with complicated systems or undetermined environmental conditions. This book chapter presents the PDEM and its applications in dynamic reliability assessment of machines and structures. The PDEM equation uses some basic concepts of probability to estimate the time-varying response of structural and mechanical systems, which can be used for accurate reliability analysis. The PDEM equation and the step-by-step procedures for dynamic reliability assessment are explained. The PDEM-based Reliability Assessment Method (PRAM) is presented in two perspectives that are offline and online PRAM. The offline PRAM is suitable for predicting the reliability in the future that is useful in the design phase for improving the design of a structure or machine based on the reliability requirements in the future. The online PRAM is suitable for the evaluation of the reliability using online monitoring data, which is beneficial for updating the maintenance policy of the system based on accurate reliability estimation. A bearing and a cantilevered beam are used as two case studies for illustrating the applicability and the advantages of PRAM for dynamic reliability assessment.

Keywords Dynamic reliability assessment · Probability density evolution · Time-varying reliability analysis · Uncertainties · Machine learning

S. Saraygord Afshari · X. Liang (✉)
Department of Mechanical Engineering, University of Manitoba, Winnipeg, MB R3T 5V6, Canada
e-mail: Xihui.Liang@umanitoba.ca

M. J. Zuo
Department of Mechanical Engineering, University of Alberta, Edmonton, AB T6G 19, Canada

F. Chaari et al. (eds.), *Smart Monitoring of Rotating Machinery for Industry 4.0*, Applied Condition Monitoring 19, https://doi.org/10.1007/978-3-030-79519-1_5

1 Introduction

Most engineering systems are comprised of structural and mechanical components that are facing different uncertainties due to the inherent randomness in both dynamic loadings and structural parameters. Reliability analysis is a practical engineering idea to assess the uncertainties associated with these systems for further increasing the reliability and reducing risks. Reliability is defined as "the probability of a system or component, performing its intended functions under specified operating conditions for a specified period of time" [1]. Many reliability assessment methods have been developed to calculate the probability of failure in a structural or mechanical system. Monte Carlos Simulation (MCS), response surface methods (RSM), and first/second order reliability methods (FORM/SORM) are some of the most common reliability assessment methods.

Two significant factors affect the accuracy of the reliability estimation. The first factor is the time-varying operating condition. For example, the rotating speed of a mechanical component such as a bearing may change versus time, and different bearings do not experience the same speed profile even when used in the same location. The second factor is the degradation of a structure or machinery due to ageing, corrosion, fatigue, etc. Hence, it is necessary to consider the time-varying condition and deterioration of structural and mechanical systems for reliability assessment. By extending the reliability calculation to time-variant machines and structures, new considerations must be taken: (1) for reliability assessment of time-variant structures and machines, the failure criteria at different points of time may change because of the change of dominant failure modes; (2) for time-variant systems, the probability of system parameters can change as well as the probability of the system response [2]. Therefore, the common reliability assessment methods become less applicable for time-variant or dynamic systems [3, 4]. In other words, the method of estimating reliability using probability density function (PDF) of limit state functions paves the road for the probabilistic design of structures and machines. However, it is not accurate enough for a time-varying reliability assessment, hence raising the need to develop dynamic reliability assessment methods.

Generally, the consideration of randomness in a given system is referred to as probabilistic system analysis, while the dynamic stochastic analysis describes the future state of the system using its history plus probabilities of successive changes [5]. Modeling of dynamic systems has been historically initiated by Einstein's studies on the Brownian motion [6], where he developed an evolution equation for particles' density suspended in a fluid and stated it as a diffusion equation. His thoughts were subsequently boosted by other scientists, including Fokker (1914), Planck (1917), and Kolmogorov (1931). As a result of these endeavors, random vibration theory became a highlighted division of stochastic dynamic analysis in the 1950s [6]. Subsequent theories and analytical techniques in dealing with dynamic linear systems were established at the turn of the century and reached a proper maturity level for application in mechanical and civil engineering fields [7].

Even with all the above progress in the stochastic analysis of linear systems, there were still significant challenges for precise response prediction of nonlinear stochastic dynamical systems. In 2004, Li and Chen presented a *Probability Density Evolution Method* (PDEM) according to the principle of preservation of probability [8], which allowed for randomness propagations in dynamical systems. The method of PDEM is of great importance and has been applied in different engineering investigations, such as the prediction of the behavior of a structure or machine or the reliability assessment of different engineering systems [9–24]. Besides the applicability of the PDEM for the design and analysis of nonlinear stochastic systems, regarding its substantial capability to predict the systems' behavior in the future under uncertain environment and loading, the PDEM has received significant attention for reliability assessment of mechanical and structural systems.

One of the most common uses of the PDEM-based reliability assessment method is the dynamic reliability analysis of nonlinear stochastic structures. In this respect, Chen et al. [25] applied the PDEM to evaluate the instantaneous PDF of the response of a general multi-DOF nonlinear dynamic structure. The reliability is then calculated through a simple integration over the safe domain. In [26], the PDEM is applied to estimate the dynamic reliability of an 8-story frame with random parameters, and it is shown that the method is efficient and accurate compared to other methods from the literature. The PDEM is also widely used for reliability assessment of structures under uncertain excitations such as seismic loading [9, 27–30]. Several studies have also used the PDEM for fatigue life reliability assessment of structures [2, 17, 31]. In general, when there exists a challenge with nonlinearity, uncertainty, or complexity of a structure, the PDEM has usually outperformed other existing methods; some examples of such situations are studied in [25, 32–35].

Not only The PDEM has been widely used for reliability assessment of structural systems, but also it has been utilized for reliability assessment and performance prediction of mechanical systems such as bearings, gearboxes, and engines. Yang et al. [36] used the PDEM to calculate the anomaly distribution of an aero-engine turbine with an initial crack. They demonstrated that the PDEM outperforms the Monte Carlo simulation for their case study. In [37] a wavelet-based PDEM is applied for reliability analysis of wind turbines; the use of PDEM for failure analysis of wind turbines is also investigated in [38]. In [39], random gust is considered using the PDEM to study the response of a stall flutter system. In [12] the PDEM is used to calculate the lifetime reliability of an aircraft wing under different damage scenarios. In [40] the PDEM is utilized to assess the performance of power systems via the calculation of dynamic probabilistic load flow. It should be noted that the PDEM equation is mathematics-based, and it can be used for a variety of other applications such as bistable systems driven by colored noise and Gaussian white noise [33], reliability of data storage of a gyroscope [41], or reliability-based active control of dynamic systems [13].

Considering the literature, the PDEM is an excellent method for performance prediction and dynamic reliability analysis of different dynamic systems. To be more specific, when facing systems with uncertain properties or considerable changes in the system or structural properties (for example, degrading systems), or systems with high levels of nonlinearity or systems that are operating under time-varying conditions, the PDEM is a constructive and practical approach. The PDEM is shown to be more effective than the Monte Carlo simulation and the subset simulation in many cases and is a feasible method to deal with the reliability analysis of complex problems [42]. This method has also been verified experimentally in several studies, such as [12, 43]. To summarize, the key advantages of using the PDEM for reliability assessment of dynamic systems are as follows: (1) The solution of the PDEM equation is a time-varying joint PDF of the system response and structural parameters. This time-varying PDF can considerably enhance the accuracy of reliability assessment for dynamic systems. (2) In most cases, the computational effort of the PDEM is considerably low compared to some other reliability assessment methods such as Monte Carlo simulation and different surrogate modeling techniques. (3) Unlike other probability density evolution equations, such as the Liouville equation and Fokker–Planck equation, this method leads to a family of equations, which are numerically amenable [44]. (4) Uncertainty and nonlinearity are decoupled in the PDEM equation, making the PDEM a simple and straightforward approach that can be used together with many other techniques to cover a wide range of reliability analysis problems. For example, online condition monitoring data can directly be fed back to the PDEM equation to update the calculated reliability and increase the accuracy of the estimated reliability.

In this book chapter, the PDEM equation is presented, and it has been explained through its physical interpretation and its application in the reliability assessment of engineering systems. Here, the PDEM-based Reliability Assessment Method (PRAM) is presented in two viewpoints that are offline and online PRAM. Offline PRAM is suitable for predicting reliability in the future, the offline term stands for the situation that we do not need real-time feedback from the system condition, and we use the initial system model to calculate the dynamic reliability of a system that is useful in the design phase for improving the design of a structure or machine based on the reliability requirements in the future. The online PRAM is appropriate for the reliability analysis using online monitoring data; in other words, the PDEM equation in online PRAM is constructed using real-time data. Online PRAM is beneficial for updating the maintenance policy of the system based on accurate reliability estimation. Followed by the introduction of the PRAM, a cantilevered beam and a bearing are used in this book chapter as two case studies for illustrating the applicability and the advantages of the PRAM for dynamic reliability assessment of engineering systems.

The organization of this book chapter is as follows: the PDEM-based reliability assessment method is described in Sect. 2. Section 3 is dedicated to the dynamic reliability assessment of structures. The application of the PDEM for dynamic reliability assessment of machines is presented in Sect. 4, followed by the discussion and future research directions in Sect. 5.

2 The Probability Density Evolution Method

In this chapter, the probability density evolution method (PDEM) is described. PDEM is a method for analyzing the evolution of the probability densities of engineering systems. It is developed based on the principle of preservation of probability. The PDEM equation is introduced in Sect. 1.1, and some notes and physical interpretations around the PDEM are given in Sect. 1.2, and in Sect. 1.3, we will explain how the PDEM can be used for dynamic reliability assessment of engineering systems.

2.1 The PDEM Equation

Generally, the PDEM equation can be derived for any dynamic system based on the system's equation of motion. The dynamic behavior of a general multi-degree-of-freedom (MDOF) system can be stated as the following equation [6]:

$$\mathbf{M}(\Theta)\ddot{\mathbf{X}}(t) + \mathbf{C}(\Theta)\dot{\mathbf{X}}(t) + \mathbf{f}(\Theta, \mathbf{X}) = \mathbf{F}(\Theta, t) \quad (1)$$

where at a given time t,$\ddot{\mathbf{X}}$,$\dot{\mathbf{X}}$, and $\mathbf{X}$ represent the acceleration, velocity, and displacement vector, respectively (they are all of the N order, which is equal to the degree of freedom), $\mathbf{F}$ is a random or deterministic external excitation, Θ is a vector of all system parameters with known (or assumed) PDF reflecting the uncertainty in excitation or physical properties. This assumed uncertainty could be in the material intensity, applied force's frequency and magnitude, or even in the type of the external force function or any other kind of other properties of the structure such as material density [45]. $\mathbf{M}$ and $\mathbf{C}$ are mass, and damping matrices of the system, respectively, and $\mathbf{f}(\cdot)$ is the restoring (elastic) force vector (for a linear system, it can be written as $\mathbf{K}_{\boldsymbol{\Theta}}(\boldsymbol{\theta}_{\mathrm{q}})\,\mathbf{X}$ where $\mathbf{K}$ is the stiffness matrix).

If we assume a fixed vector of Θ, Eq. (1) becomes a deterministic equation and if we assume a PDF for the vector of random variables, Θ, Eq. (1), turns into a probabilistic equation that can be solved through different simulation schemes such as MCS and importance sampling. Nevertheless, when considering the past plus changes in the probability of Θ versus time, we will face a stochastic equation. To solve the equations of motion in a stochastic form, we need to set up the relationship between the initial PDF of the system's random factors and the instantaneous PDFs of the same parameters at the desired time interval [46]. That relationship can be concluded from the principle of the preservation of probability. Based on the principle of the preservation of probability [46], if no new random factors appear and the existing random factors do not disappear, the probability within the system will be preserved [47]. The mathematical formulation of the principle of preservation of probability and other necessary considerations for using that principle are provided in [46]. Now, based on the principle of preservation of probability, the PDEM equation for an MDOF system as given in Eq. (1), can be written as:

$$\frac{\partial P_{\mathbf{X}\Theta}(\mathbf{X}, \theta, t)}{\partial t} + \sum_{l=1}^{m} \mathbf{h}_{\mathbf{X},l}(\theta, t) \frac{\partial P_{\mathbf{X}\Theta}(\mathbf{X}, \theta, t)}{\partial \mathbf{X}_l} = 0 \tag{2}$$

Equation (2) is called the general evolution equation for calculating the joint probability density of the specified dynamic response of a system with specified system parameters. In this equation, m represents the number of dimensions, $\mathbf{X}_l$ is the system response vector in the specified direction, and $\mathbf{h}_{\mathbf{X},l}(\theta, t)$ is the derivative of the response vector. It should be noted that when a one-dimensional physical quantity is of interest (e.g., mass or displacement in a specific direction), $m = 1$, and the probability evolution equation reduces to:

$$\frac{\partial P_{\mathbf{X}\Theta}(\mathbf{X}, \theta, t)}{\partial t} + \mathbf{h}_{\mathbf{X}}(\theta, t) \frac{\partial P_{\mathbf{X}\Theta}(\mathbf{X}, \theta, t)}{\partial \mathbf{X}} = 0 \tag{3}$$

or in a more straightforward form:

$$\frac{\partial P_{\mathbf{X}\Theta}(\mathbf{X}, \theta, t)}{\partial t} + \dot{\mathbf{X}} \frac{\partial P_{\mathbf{X}\Theta}(\mathbf{X}, \theta, t)}{\partial \mathbf{X}} = 0. \tag{4}$$

The initial condition is:

$$\partial P_{\mathbf{X}\Theta}(\mathbf{X}, \theta, t)|_{t=t_0} = \delta(\mathbf{X} - \mathbf{X}_0) p_{\Theta}(\theta) \tag{5}$$

where $\mathbf{X_0}$ is the initial vector of the physical response variable. Now, using the finite difference method, the PDEM partial differential equation (Eq. 4) can be solved together with the physical equation (Eq. 1). In other words, to solve the PDEM equation (Eq. 4), we need to find $\dot{\mathbf{X}}$ and substititue it into the PDEM equation (Eq. 4), then using the introduced initial condition (Eq. 5), the joint PDF of the system response and uncertain system parameters, θ will be concluded. it should be noted that there is no limitation in the means to find $\dot{\mathbf{X}}$, it can be calculated via solving the system's equations of motion and simulating the system response or it can be directly fed back to the PDEM equation from a condition monitoring system.

The detailed steps to calculate the PDEM equation using the principle of the preservation of probability can be found in [6, 45].

2.2 *Physical Interpretation of the PDEM*

The probability density evolution method was developed from physics-based dynamic equations, and, as noted before, its theory is based on the probability preservation principle. There are other conservation laws in nature, such as the conservation of mass and the conservation of energy. The probability of a random event is also conserved, which is the perspective of the principle of probability conservation. Li and Chen [47] explained that principle from the perspective of random event description [48]. The critical assumption for using the probability preservation principle is that the existing random factors do not disappear, and no new random factor is added to the random system. If we go far back through the physical interpretation of the principle of preservation of probability, this principle, and its key assumptions emanate from the first law of thermodynamics, known as the Law of Conservation of Energy. The law of conservation of energy states that "energy can neither be created nor destroyed; energy can only be transferred or changed from one form to another" [49].

From the PDEM equation and due to the randomness propagation, the probability density function (PDF) of any random event will be different at any time in the processes of evolution. The change of the PDF will result in a specific probabilistic response of the system at any specific point in time. The significant difference between the PDEM method and other reliability calculations methods is that the PDF of system response is not a constant function, and it can change versus time based on all random factors associated with the system and environment. The other methods that can account for randomness propagation in physical systems are methods such as the Fokker-Plank equation [50], which is very hard to solve for real systems and structures.

2.3 Dynamic Reliability Assessment Using PDEM

Generally, assuming a safe domain of system response, Ω_s, and having the probability of system response at the time τ, as $P(\mathbf{X}(\tau))$, the system reliability can be stated as:

$$R(t) = P\{\mathbf{X}(\tau) \in \Omega_s;\ \tau \in [0, t]\} \tag{6}$$

Equation (6) indicates that the reliability is related to the probability of the random event's response stating in the safe operation domain over the operational time interval [0, t]. Here, Ω_S is directly related to the definition of the failure criteria by a designer or an operator, and it can be concluded from primary numerical analyses or tests. For example, in a structural system, Ω_S represents the maximum allowed displacement that is the failure threshold, or in a mechanical component such as a bearing or gearbox, Ω_S denotes the maximum allowed vibration amplitude.

Upon solving the PDEM equation, the time-varying joint probability density of the system response and uncertain parameters, $p_{\mathbf{X}\Theta}(\mathbf{X}, \theta, t)$, will be calculated. "For example, if the system response of a structure is displacement, and the uncertain parameter is structural stiffness, the joint probability density of the displacement and the structural stiffness can be calculated via the PDEM as a function of time."

If the solution of this equation is considered as $\tilde{P}_{X\Theta}(X,\Theta)$ at time t, (denoted below as $\tilde{P}_{X\Theta}$ (X, Θ, t)), then the total PDF of structure response can be estimated by integration over the uncertain parameter's domain of variation, Ω_Θ, as:

$$\tilde{P}_X(X, t) = \int_{\Omega_\Theta} \tilde{P}_{X\Theta}(X, \Theta, t) \mathrm{d}\Theta \tag{7}$$

Finally, the equation for the PDEM-based reliability assessment method (PRAM) considering predefined failure criteria (or the safe domain) can be given by:

$$R(t) = \int_{\Omega_\Theta} \tilde{P}_X(X, t) \mathrm{d}X \tag{8}$$

In other words, Eq. (8) integrates over the probable structural responses before reaching the failure state (Ω_s boundaries). As it is evident from Eq. (8), the estimated reliability via the PRAM is not a constant value anymore, and it is a function of time, which is more suitable for assessing the reliability of a dynamic system. The complete procedure of using the aforementioned steps towards estimating reliability through the PRAM is presented in the flowchart of Fig. 1. Regarding this flowchart, it can be seen that the user needs to find a system model at first, based on a measured system

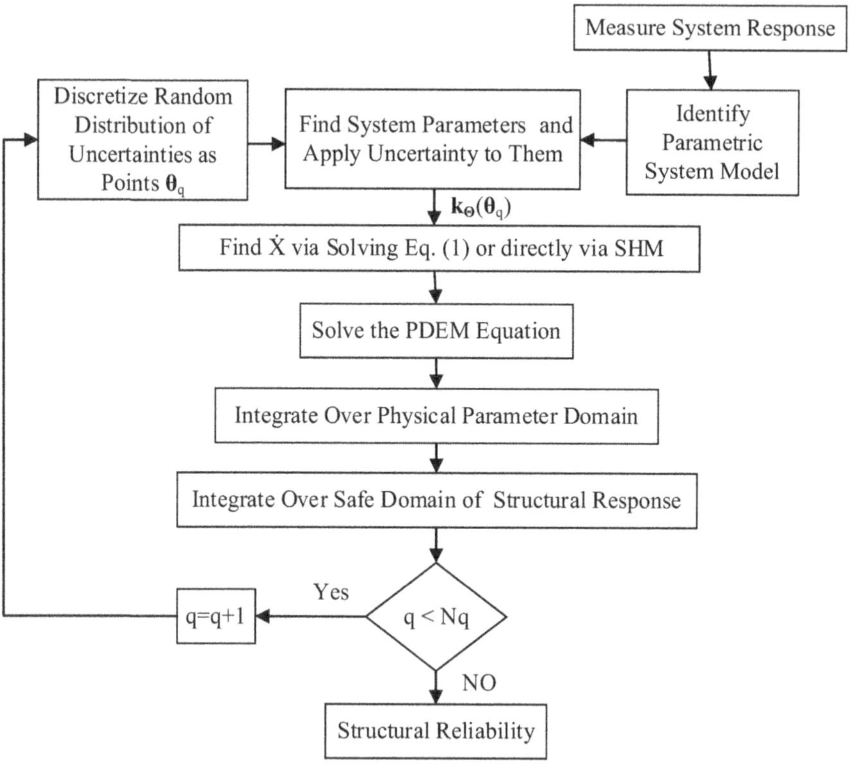

Fig. 1 Proposed algorithm for the PDEM-based reliability assessment method

response or an analytical method. Then, the user has to assume a probability density for the model parameter (s), to generate discretized parameters to be substituted to the system equation, Eq. (1), for further calculation of the system response and also solving the PDEM equation (Eq. 3). The PDEM results in a joint probability density function of the system response and the uncertain parameter ($\tilde{P}_{X\Theta}$ ($\boldsymbol{X}$, Θ, t)). After integrating over the physical parameter's domain of variation, a time-varying probability density function for the system response will be calculated, which can be used for the calculation of the reliability, as in Eq. (8).

Afterward, in Sects. 3 and 4, two examples are presented to demonstrate the applications of the PDEM. The first example (Sect. 3) is the application of PDEM for structural reliability analysis, and the second example (Sect. 4) is dedicated to the dynamic reliability assessment of a bearing.

3 Dynamic Reliability Assessment of Structures

One of the most useful applications of stochastic response analysis of mechanical and structural systems via the PDEM is to make a foundation to improve the safety, reliability, and serviceability of engineering systems in their expected service life. As mentioned before, Reliability is defined as "the probability of a system or component, performing its intended functions under specified operating conditions for a specified period of time." Hence, an accurate reliability estimation depends on the definition of three important factors: (1) intended function (which also results in the definition of failure criteria), (2) operating condition, and (3) the time that is connected to the service life. For a dynamic structure, the factor mentioned above can change versus time. Thus the reliability can also change.

Regarding that background, the main objective of reliability analysis is to estimate the probability of the system response not exceeding the failure criteria. For a dynamic reliability assessment, the dynamic system response can be concluded from the PDEM. In this section, two approaches for using the PRAM are introduced, and a cantilevered beam is also experimentally tested as a case study to demonstrate the applicability of the PRAM for dynamic reliability assessment of structures.

3.1 *Offline PDEM-Based Reliability Assessment Method*

There are two key steps towards performing the PRAM that are constructing the PDEM equation (Eq. 4) and solving the PDEM equation. Comprehensive explanations about the method of solving the PDEM equation are provided in [19, 51]. Here in this chapter, we are introducing two different approaches for constructing the PDEM equation that are the offline and online PDEM methods.

In order to construct the PDEM equation, we need to substitute into Eq. (4). If we are in the design phase, or we are not able to run the system for a long time to monitor and collect $\dot{\mathbf{X}}$ values for all the expected service life, we can solve the system equation of motion to simulate $\dot{\mathbf{X}}$ and then substitute it into the PDEM equation (Eq. 4). Here we call this method "offline PRAM" because we do not require real-time feedback from the system condition to estimate the reliability. Another alternative for assembling and solving the equations of motion is using a system identification technique to identify the system response function. The latter technique can be applied to both linear and nonlinear systems. There are various methods for system identification, such as classical regression methods and neural networks, mostly used for nonlinear systems.

3.1.1 Application of the Offline PRAM

In many systems with high safety and reliability demands, the design is optimized based on the reliability requirements. For example, imagine a satellite system that is supposed to be fully functional for twenty years in the space. It is not feasible to use high safety factors when designing that satellite, because a large safety factor burdens higher weight to the system and probably makes that satellite too heavy for being launched through a launch vehicle. Therefore, systems like that exemplary satellite are usually designed based on their service life reliability. On the other hand, for a system that is going to be used for many years, it is not feasible to perform a real lifetime test in order to collect the lifetime data and calculate the system reliability. In such cases, using the offline PRAM would help engineers and designers to use a simulation scheme for finding the dynamic reliability of the system for its service life.

3.2 *Online PDEM-Based Reliability Assessment Method*

As explained in the previous section, we must calculate or monitor the $\dot{\mathbf{X}}$ value in order to substitute it into Eq. (4) to perform the PRAM. It is explained that how $\dot{\mathbf{X}}$ can be calculated with a computer simulation to achieve an offline reliability estimation via the PRAM. The offline PRAM provides a potent tool for design optimization and safety analysis of engineering systems. However, when online monitoring data from the current state of structural and mechanical systems are available, the accuracy of the PRAM can be further improved by directly substituting $\dot{\mathbf{X}}$ from the condition monitoring unit into Eq. (4). Here, the latter technique is called online PRAM.

3.2.1 Application of the Online PRAM

With the help of modern condition monitoring systems, gathering the information of actual system response has become feasible. This information can be used to improve reliability estimation. Based on an accurate and near-to-actual reliability value, operational staff can be informed ahead of any critical operation state. For a large engineering system with numerous set of components, maintenance scheduling is a critical task which can affect many factors such as safety and cost. An accurate online reliability assessment method such as online PRAM would greatly help to maintain a system with an optimum cost and appropriate safety.

In the next section, we will perform both offline and online PRAM for a structural system, which is a composite piezo-laminated cantilevered beam.

3.3 Case Study: Cantilevered Beam

In this book chapter, the PDEM-based reliability assessment method (PRAM) is introduced in two different manners. One of the most common applications of the PRAM is the reliability assessment of structural systems. Here, in this section, the application and accuracy of the PRAM for reliability assessment of a cantilevered beam structure is experimentally investigated. Cantilevered beam structure is one of the simplest and most used structural components in engineering systems; hence it can be an excellent example for verifying the PRAM applicability for structural reliability assessment. The experimental test and the presented case study are previously studied by Afshari et al. [13], and it is also presented here for a better understanding of the PDEM-based reliability assessment method.

3.3.1 Experimental Setup

As presented in Fig. 2, a cantilevered composite ($E = 20$ GPa, $\rho = 1197$ kg/m^3), is used for the experimental study of the online and offline PRAM. Two piezoelectric actuators are bonded on both sides of the composite beam for applying the necessary actuation force as an external loading. A piezo-sensor is also attached at the fixed end of the beam for strain acceleration sensing. The thicknesses of the beam and piezoelectric patches are 2 mm and 0.6 mm, respectively. Additional details about the dimensions are provided in Fig. 3, and a schematic diagram of the experiment is shown in Fig. 4. In this experiment, all piezo-patches are surface-bonded at the fixed end of the beam. A PIEZO SYSTEM INC. 20X amplifier is used to excite the piezo-actuator, and PicoScope® 5000 Series data logger is utilized as an analog–digital converter.

3.3.2 System Identification

In Sect. 3.1, it is explained that in order to perform an offline PRAM, first, a mathematical model of the system is required. The mathematical model will be used to simulate the system response during its expected service life for further construction of the PDEM equation. Here, the frequency response function (FRF) of the

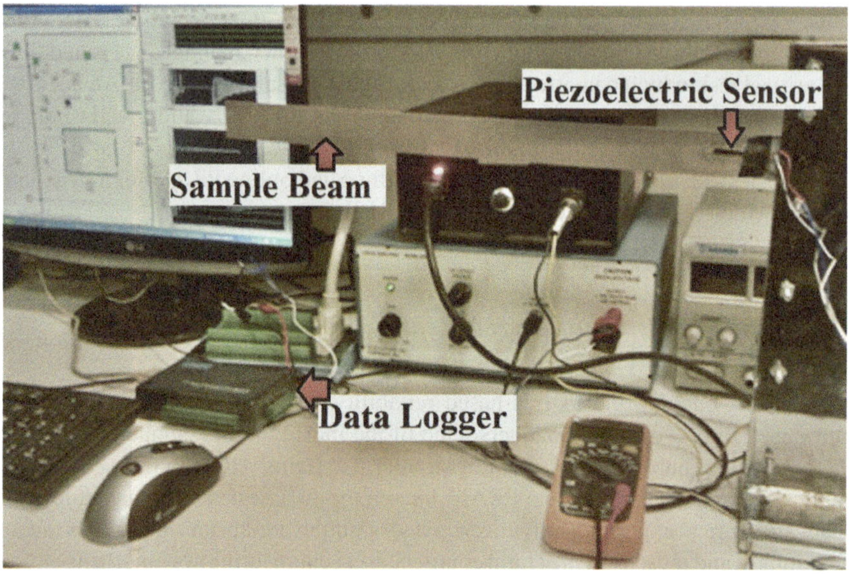

Fig. 2 Experimental setup: a cantilevered beam structure with bonded piezoceramic sensors and the actuation system

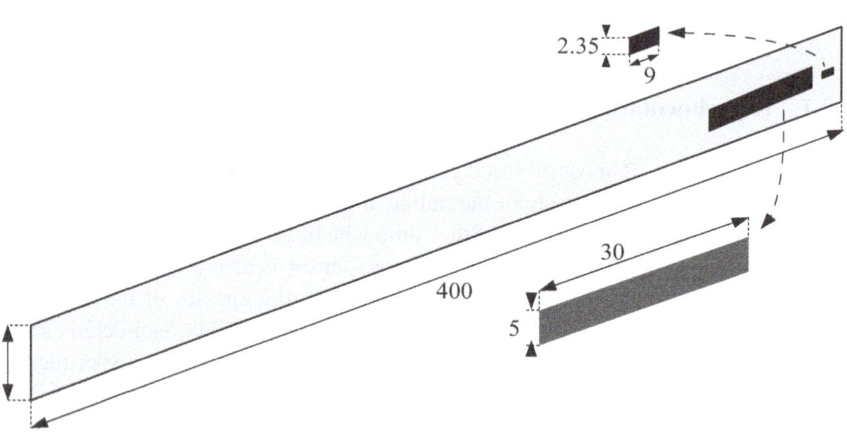

Fig. 3 The beam schematic and its dimensions (mm) [14]

cantilevered beam is used to extract a numerical model for the cantilevered beam. The utilization of the FRF to estimate a system model is known as nonparametric system identification, and it is a common approach in structural systems [52].

Here, FRF data is obtained by applying a sinusoidal sweep excitation to the beam to excite the natural modes of the beam within the desired frequency domain. The excitation is performed by exciting the piezo-actuators. The piezoceramic sensor measures the resulting strain-induced voltage. The sampling frequency is 1000 Hz

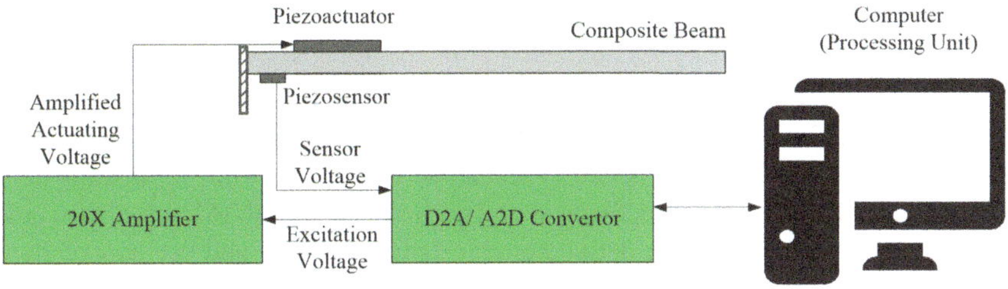

Fig. 4 Schematic of experimental reliability testing system [14]

and a digital low-pass filter to avoid capturing unnecessary signals. The calculated FRF is demonstrated in Fig. 5 using the black curve, which presents the magnitude of system response as a function of frequency.

The next step is to find a parametric (mathematical) system model that can be extracted from the calculated FRF. Here, using the procedures outlined in [53], a suitable parametric system representation is identified via the so-called "prediction error method (PEM)." In this regard, we have tried different model orders (from order 6 to order 25) to see which model complies the best with the real data in their frequency response function. The best model order to fit the real response is found to be of order 8; therefore, the final parametric model to be used is an order-8 model. The final mathematical parametric model is presented in Fig. 4, using the red curve. For example, for a similar cantilevered beam structure, an analytical solution has been presented in [53], and it is shown that both models are performing well for simulating the system response.

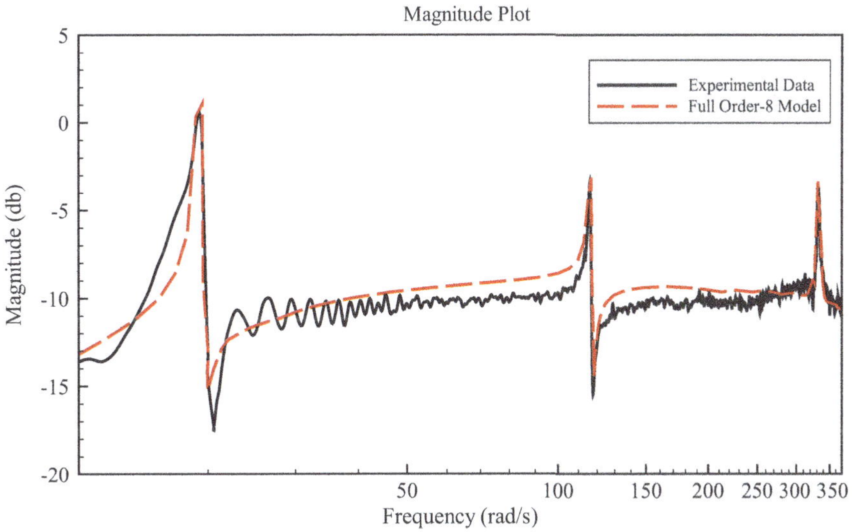

Fig. 5 Comparison between experimental FRF and a parametric model [14]

3.3.3 Accelerated Reliability Test

In this study, in order to validate the results of the PRAM, accelerated reliability tests are also carried out to calculate the real reliability of the beam structure experimentally. In this regard, a test is designed to accelerate the failures while replicating field performance. To assess the effectiveness and accuracy of the PRAM, it is also necessary to apply proper loadings, define suitable failure criteria, and use an excellent progressive damage type. Here, to make progressive damage type, artificial delamination was embedded in the test samples by implementing a waxed thin plate amid the plies during the process of structure prototyping. In the presence of artificial damage, the structural stiffness will decrease, and this reduction should be manifested in the structural response. Here, with the purpose of evaluating structural strength degradation during repeated external loadings, a sinusoidal excitation force with random frequency and intensity is applied to the damaged structure for a time period of 120 s. Depending on the uncertainties associated with the physical parameters and the excitation force, and the predefined failure criteria, the structure might fail anytime during the test. For an accelerated test, the failure criteria must be chosen appropriately to define failures during the specified time of the experiment. It has been proven that the output of attached piezoelectric sensors during the beam excitation is directly related to the yield stress at the cantilevered end of the beam [53, 54]. Considering this point, here, pre-experimental tests and analyses were taken first, and they have revealed that when the piezoelectric sensor output is more than 3.2 V, the system should be considered "failed" because the stress at the end of the beam will be more than the yield stress.

3.3.4 Offline and Online PRAM for the Cantilevered Beam

The PRAM approach for reliability assessment has been described in Sect. 3. Here, both offline and online PRAM have been applied for the beam structure to evaluate the effectiveness of PRAM via a comparison with experimental reliability results. In order to perform the offline PRAM method on the introduced beam, first, a normal distribution has been assumed for the structural stiffness and the random excitation. Using those assumptions together with Eq. (1), the system response can be concluded versus time. The calculated response is substituted in Eq. (4) to estimate the joint PDF of the response and structural stiffness. In order to solve the PDEM equation, the initial condition is assumed as in Eq. (5) where a normal distribution for the uncertain parameter, $p_{\Theta}(\theta)$, (here θ the identified stiffness) is assumed with the coefficient of variation of 10%, this assumption depends on the accuracy of the measurement and modeling process. The estimated joint PDF is then used in Eqs. 7 and 8 to estimate the dynamic reliability of the beam versus time.

For the online PRAM, an updated $\dot{\mathbf{X}}$, that is obtained from a condition monitoring sensor has been used in the PDEM equation (Eq. 4) to upgrade the accuracy of the reliability evaluation. An evolution of the PDF of the system response estimated via the PDEM equation is presented in Fig. 6. Figure 7 presents the results of both

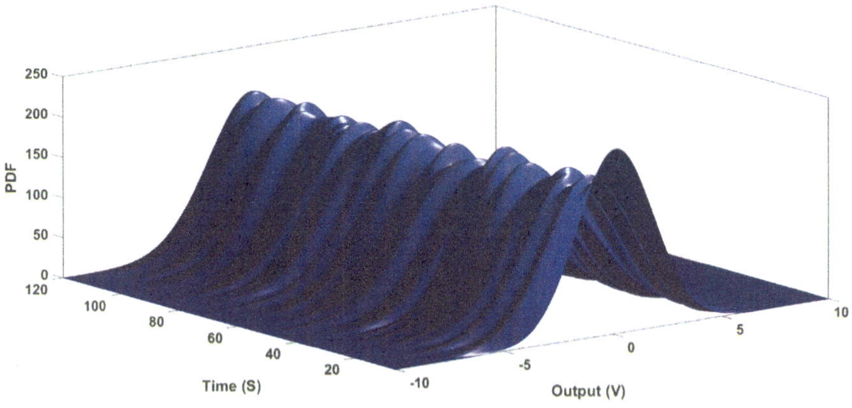

Fig. 6 Evolution of the probability versus time

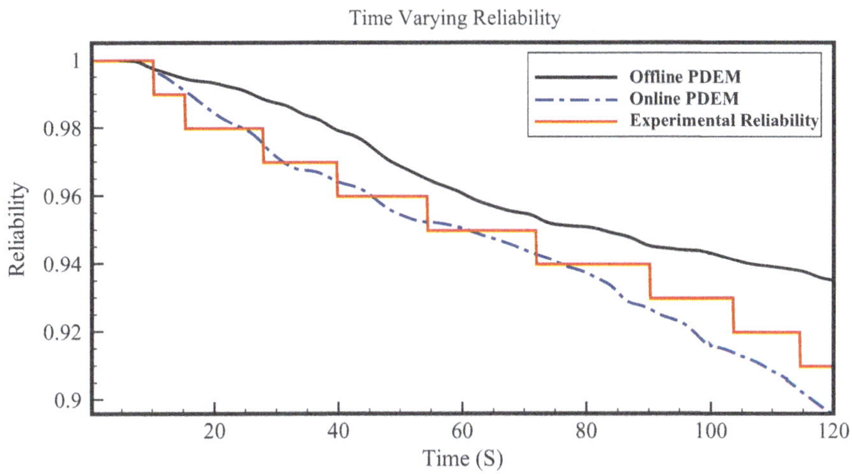

Fig. 7 Dynamic reliability curves using different approaches [14]

offline and online PRAM for reliability assessment of the beam as compared with experimental reliability results. It is realized that both offline and online methods result in similar trends; however, the online PRAM curve is closer to the experimental reliability curve. The procedure for calculating the experimental reliability curve is discussed in the next section.

3.3.5 Experimental Dynamic Reliability of the Beam

Aiming to experimentally evaluate the accuracy of the PRAM, more than a hundred duplicate samples were made. All samples were exposed to the same damage and loading condition in a similar environment, and the experimental time responses of all samples were gathered for 120 s. For the calculation of the experimental reliability curve, each time when the response of the sample under excitation crosses the failure criteria, it decreases the reliability value at that point in time. The calculated reliability that is calculated using the reliability evaluation tests is also imported in Fig. 7.

It should be noted that if we increase the number of samples in the reliability tests, a smoother experimental curve can be established [13]. However, from the trends, it is evident that the PRAM performs with acceptable accuracy in this experiment, and the online PRAM is more accurate than the offline PRAM. One major reason for the less accuracy of the offline PRAM is the probable variations and sudden changes of the material physical parameters versus time. In other words, the mathematical model used for offline PRAM is not proper for modeling sudden changes in the system. This results in an imprecise result for the offline PRAM. On the other hand, in the online PRAM, we are using real data as the input of the PDEM equation; thus, the final result of the online PRAM will be more accurate.

4 Dynamic Reliability Assessment of Machines

As discussed in the previous sections, the PDEM equation is founded on the ground of mathematical and physical concepts. Hence, the PRAM can be used for any engineering system such as structures and machines. In this section, first, some necessary considerations are explained for using the PRAM for dynamic reliability assessment of machines; then, a rolling bearing is used as a Case study to show the applicability of PRAM for reliability assessment of machinery.

4.1 Extra Considerations for Dynamic Reliability Assessment of Machines

Here, we list some considerations when using the PRAM for dynamic reliability assessment of machinery:

1) For a structural system, stiffness and/or damping is usually used as the uncertain physical parameter, Θ. However, for a bearing, several other parameters can construct the vector of Θ. In addition to the stiffness and damping, some parameters that can be used for bearings are ball diameter, contact angle, and load rating.
2) When there are several physical or performance parameters of the system that are of interest or measurable, one can take them all into the PDEM equation by defining a new variable vector, as in Eq. (9):

$$\mathbf{Z}(t) = \Psi[\boldsymbol{X}(t), \boldsymbol{Y}(t)] \tag{9}$$

where Ψ in Eq. (9) is an operator in which transforms different uncertain state variables into only one uncertain vector. Using this transformation, the state vector, $\boldsymbol{X}$, in Eq. (3) can be replaced by $\boldsymbol{Z}$. Such a transformation may result in a multidimensional PDEM equation, which can be solved via the methods such as finite-difference, which is represented in [51]. For example, in a bearing, the vector of $\boldsymbol{Z}$, can be constructed using all the data from horizontal vibration sensors, vertical vibration sensors, and temperature sensors.

3) If our focus is one failure mode or one performance parameter, it is reasonable to take the most critical failure mode or the failure criteria that results in an earlier failure.
4) In the PDEM equation, the state or displacement, $\boldsymbol{X}$, is a function of system uncertain parameters (Θ). The uncertain parameter does not necessarily need to have a direct physical meaning; in other words, they can be representative of a physical parameter. However, the PDF of Θ must have a measurable value. For example, the uncertain parameter can be the voltage output of a sensor with a normal distribution.

Considering the above explanations, in Sect. 4.2, the reliability of a rolling element bearing is investigated using offline and online PRAM.

4.2 Case Study: Bearing

To investigate the applicability of the PRAM for reliability assessment of machinery, as an example, data of accelerated degradation tests for fifteen rolling element bearings are used in this section. The same as the previous section, the data are initially used to find a mathematical model for the system to perform offline PRAM. Then using real-time data, the online PRAM is also performed, and the results are compared.

4.2.1 Data Description

Here, rolling element vibration data that are presented in [55] are used for the evaluation of the PRAM. A schematic of the experimental setup is shown in Fig. 8. Accelerated degradation tests of fifteen bearings (type: LDK UER204) are performed using this platform, and data collected as presented in Table 1. Different loading forces and frequencies have been applied to the bearings, and different failure modes took place as demonstrated in [55]. Additional details about the dataset are also presented in [55]. The uncertainties associated with the data must be known before performing the PRAM. The selection of system uncertainty characteristics depends on the system characteristics and the testing facilities as well as environmental factors. In order to find the joint probability density of the response, we need to have a distribution domain for the uncertain parameter, θ, to use in Eq. (7). This distribution can

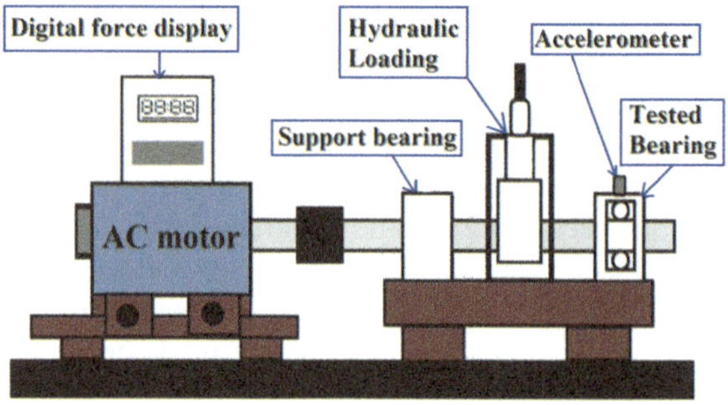

Fig. 8 Schematic of the bearing experimental testing setup [55]

Table 1 Tested bearings datasets

Loading condition	Bearing number	Bearing lifetime (min)	Fault element
35 Hz 12 kN	1–1	123	Outer race
	1–2	161	Outer race
	1–3	158	Outer race
	1–4	122	Cage
	1–5	52	Inner and Outer race
37.5 Hz 11 kN	2–1	491	Inner race
	2–2	161	Outer race
	2–3	533	Cage
	2–4	42	Outer race
	2–5	339	Outer race
40 Hz 10 kN	3–1	2538	Outer race
	3–2	2496	Inner and Outer race, ball, cage
	3–3	371	Inner race
	3–4	1515	Inner race
	3–5	114	Outer race

be concluded from the accuracy of the measurement system or the uncertainty of the system's physical parameters related to the manufacturing quality or any other possible uncertainties associated with that parameter. Here, we have used an existing dataset (the experiments were not conducted by us); therefore, we do not have enough information to estimate the parameter uncertainties. We assumed that the θ, which is the bearing contact angle, follows the normal distribution with a coefficient of variation of 10%. The coefficient of variation is to account for the possible uncertainties in the measurement/identification of the uncertain parameter. For sure, a more accurate estimation of this coefficient of variation will result in a more accurate reliability estimation. In our future work, we will improve our method by accurately estimating

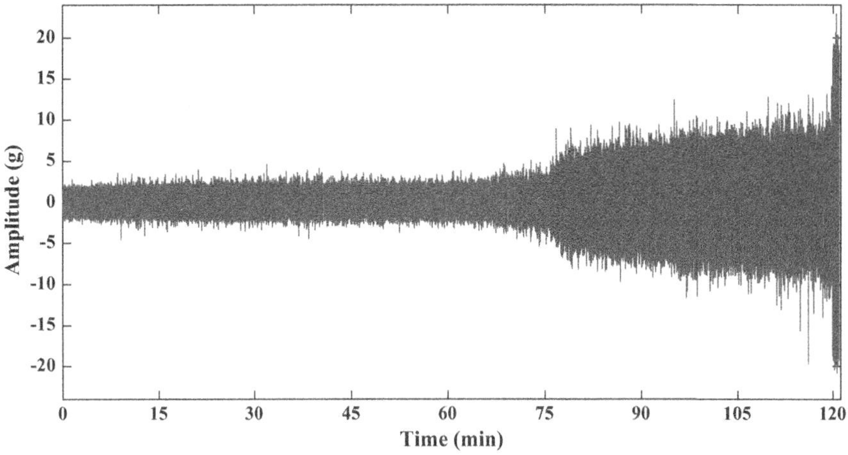

Fig. 9 Horizontal vibration signal for bearing 1–1

the parameter uncertainties. It should be noted that θ in this case study is assumed to be the bearing contact angle.

As depicted in Fig. 9, the vibration increase versus time in the phase of degradation. This shows that the vibration amplitude can be taken as a failure criterion [56]. Wang et al. have taken the vibration amplitude as the failure criteria in their research [55]. Here, based on the presented PRAM method, we will estimate the dynamic reliability of the bearings using the same failure criteria as in [55].

4.2.2 System Identification

Similar to the structural reliability assessment, in order to perform the offline PRAM for dynamic reliability assessment of rolling element bearings, a system model is needed to simulate the response. For the bearing data, there is no sweep frequency excitation data, but as presented in Table 1, data for 15 excitations with three different loading and frequencies are available. Thus, we can use a multi-input-multi-output (MIMO) system identification method to find a proper mathematical model for the rolling element bearing system. For this purpose, we use the first 50 min of all time responses as the output of the MIMO system identification technique.

Here, a MIMO neural network is used to estimate the bearing model, using the time response data listed in Table I. The bearing speed and load are used as the input of the network, and the vibration amplitude is used as the output. Therefore the model can be used to predict the system response based on the input speed and load. In order to find a mathematical model of the bearing, we split the experimental data of the bearings time responses into three sets, 9 for the training, 3 for the verification, and 3 for the test in the BPNN model. Ten hidden layers with 15 neurons in each layer established the model. It was found that 10,000 epochs can be considered as an

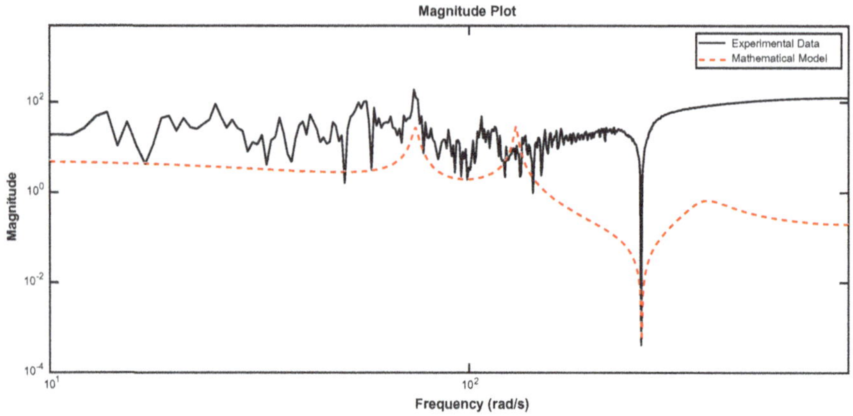

Fig. 10 The frequency response of the rolling element bearing

adequate number for the training process. The frequency response calculated using the experimental data and also the frequency response of the mathematical model are both illustrated in Fig. 10. Although the FRF of the mathematical model is not accurately matching with the experimental FRF, there is a proper matching at the frequencies that we are going to simulate the system numerically. As mentioned before, the uncertainty assumptions will also account for some possible errors in the mathematical model. Now, as the next step, the identified system can be utilized with Eq. 4 in order to solve the PDEM equation to perform the offline PRAM.

4.2.3 Offline and Online PRAM for the Rolling Bearing

Here, both offline and online PRAM have been applied to investigate the applicability of the presented method for reliability assessment of a bearing. To perform the PRAM, a normal distribution for both the bearing vibration signal and the excitation has been presumed. All the steps for offline and online PRAM are the same as the steps carried out for structural reliability. The only difference is that here, the bearing contact angle (initially $= 0°$) is used as the uncertain physical parameters instead of the structural stiffness. It is worth mentioning that wear is one of the reported failure types in this set of experiments, and the changes of contact angle are related to the wear as a failure mode [57]. Here, initially, we assumed a normal distribution for the contact angle. The updated values of the contact angle can be updated versus time using the updated system response. Detailed mathematical relations between these parameters and bearing vibration response are explained in [58].

Figure 11 presents the results of the two methods. The experimental reliability curve is also included in Fig. 11. The experimental curve is calculated using 5 datasets that are collected for the same bearing under the same working condition (loading characteristics: 35 Hz and 12 kN). For example, in the experimental curve, at t

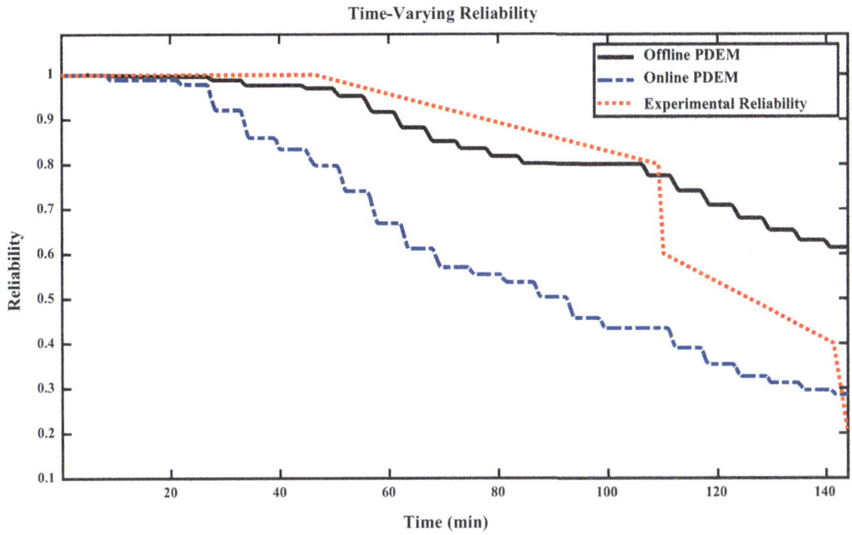

Fig. 11 Comparison of the time-varying reliability curves via different approaches

= 113 min, the reliability value is 0.6, which means that amongst all five tested bearings, two of the bearings have failed before $t = 113$ min. It is worth recalling that the existing experimental data for the investigated bearing are not sufficient to plot a trustworthy experimental reliability curve, and we just provided this curve here to make a rough comparison. Our investigation on the problem of bearing reliability is still preliminary work, and it is only used as an example of the PRAM application for machinery. This study will be further improved in the future after performing more experimental investigations.

5 Discussion and Future Research Directions

In this chapter, the probability density evolution method (PDEM) is presented to evaluate the dynamic reliability of a structure or a machine. It is shown that the PDEM equation has proper flexibility, and it can be applied in different frameworks that are offline and online PDEM-based reliability assessment (PRAM). However, the accuracy of the PRAM is strongly dependant on the following assumptions or inputs: failure criteria, the accuracy of measured or simulated $\dot{\mathbf{X}}$, the initially assumed PDFs for the system uncertain parameters, and uncertain parameters to be included in the vector of Θ. Also, for the offline PRAM, it is crucial that the system does not go under a sudden or rapid change.

For example, from Fig. 11, it can be realized that both offline and online reliability curves for the bearings are representing a pretty similar trend. However, as compared to the dynamic reliability curves for a cantilevered beam (Fig. 7), the difference between the offline and online curves is more significant for the rolling element bearings. It is the occurrence of a rapid failure in the rolling element bearing that makes the offline prediction to be less accurate. This shortcoming in the use of PDEM for systems with rapid or sudden failures can open a new direction for future studies in the field of dynamic reliability analysis of degrading systems via the PDEM.

The definition of proper failure criteria is also a critical point in reliability estimation. In order to define proper failure criteria, failure modes must be recognized first. For every failure mode, distinct failure criteria can be defined. Each failure criterion corresponding to a failure mode can be dependent, partially dependent, or even independent from other failure modes. For example, the failure criteria for a bearing can be defined based on its temperature, vertical vibrations, horizontal vibrations, etc. In mechanical systems, the types of failure modes are more diverse than the structural system because the failure criteria of a mechanical system can be defined based on both structural characteristics of a mechanical component and the dynamic performance of that component.

5.1 Future Research Directions

Regarding the abovementioned notes, the following research directions are suggested in the future to enhance the accuracy of the PRAM:

- For the offline PRAM, novel approaches for modeling degrading systems can be used together with the PDEM to increase the accuracy of degrading systems.
- For highly time-variant systems, some classification methods can be applied to combine correlative failure modes.
- The failure criteria can be defined as a dynamic variable, and it can be updated versus time. Another PDEM equation can also be defined to find an estimate for the dynamic failure criteria.
- The primary uncertainty distribution for parameters is also a determinative factor, and its accuracy will affect the performance of the PDEM. Further studies on improving the accuracy of uncertainty distributions can be a good area of study in PDEM applications.
- There are many uncertainties associated with a system, both in the system parameters and loading characteristics. Because of the high computational effort, it is inefficient to take all uncertainties into account while performing the PDEM. Therefore, it is determinative to study the proper set of uncertainties to be taken in the PRAM procedure.
- The PDEM is not able to model sudden failures, but it can be used together with data analysis techniques to account for sudden failures. This can be an interesting topic for future studies.
- For the systems that are highly degrading with time, different correction factors can be added to the PRAM, and it should be studied in the future.

Acknowledgements This research is supported by the Natural Science and Engineering Research Council of Canada, Canada [RGPIN-2019-05361, grant number RGPIN-2015-04897]; Future Energy Systems under Canada First Research Excellent Fund [FES-T14-P02]; and the University of Manitoba Research Grants Program (URGP). The authors would like to appreciate the work done by Prof. Yaguo Lei and the Institute of Design Science and Basic Component at Xi'an Jiaotong University (XJTU), and the Changxing Sumyoung Technology Co., Ltd. for sharing their experimental dataset of run-to-failure for rolling element bearings. The authors also acknowledge that the experiments for the cantilevered beam structure were conducted at the Composite Research Network (CRN) Lab, School of Engineering, the University of British Columbia.

References

1. Barlow RE, Proschan F (1996) Mathematical theory of reliability, vol 17. SIAM
2. Xu Y (2015) Fatigue reliability evaluation using probability density evolution method. Proba Eng Mech 42:1–6
3. Leonel ED et al (2010) Coupled reliability and boundary element model for probabilistic fatigue life assessment in mixed mode crack propagation. Int J Fatigue 32(11):1823–1834
4. Zhou Q et al (2017) Time-variant system reliability assessment by probability density evolution method. J Eng Mech 143(11):04017131
5. Kumar PR, Varaiya P (2015) Stochastic systems: estimation, identification, and adaptive control. SIAM
6. Li J (2016) Probability density evolution method: background, significance and recent developments. Probab Eng Mech 44:111–117
7. Lutes LD, Sarkani S (2004) Random vibrations: analysis of structural and mechanical systems. Butterworth-Heinemann
8. Li J, Chen J (2004) Probability density evolution method for dynamic response analysis of structures with uncertain parameters. Comput Mech 34(5):400–409
9. Zhang J, Xu YL, Li J (2011) Integrated system identification and reliability evaluation of stochastic building structures. Probab Eng Mech 26(4):528–538
10. Li J, Chen J-B, Fan W-L (2007) The equivalent extreme-value event and evaluation of the structural system reliability. Struct Saf 29(2):112–131
11. Li J, Chen JB (2006) The probability density evolution method for dynamic response analysis of non-linear stochastic structures. Int J Numer Meth Eng 65(6):882–903
12. Afshari SS, Pourtakdoust SH (2018) Probability density evolution for time-varying reliability assessment of wing structures. Aviation 22(2):45–54
13. Saraygord Afshari S, Pourtakdoust SH (2018) Utility of probability density evolution method for experimental reliability-based active vibration control. Struct Control Health Monitor **25**(8): e2199
14. Xiao NC, Zuo MJ, Zhou C (2018) A new adaptive sequential sampling method to construct surrogate models for efficient reliability analysis. Reliab Eng Syst Saf 169:330–338
15. Chen J et al (2019) Stochastic harmonic function based wind field simulation and wind-induced reliability of super high-rise buildings. Mech Syst Sign Process 133:106264
16. Gu ZY et al (2019) Dynamic reliability analysis of large span isolated structures based on extreme distribution theory. IOP Conf Ser: Earth Environ Sci 283:12041
17. Hou H-M, Dong G-H, Xu T-J (2019) Fatigue damage distribution and reliability assessment of grid mooring system for fish cage. Mar Struct 67:102640
18. Jiang G et al (2017) A storage reliability evaluation method of gyroscope based on probability density evolution. IEEE
19. Jiang G, Yuan H, Zhang H (2017) Estimating reliability of degraded system based on the probability density evolution with multi-parameter. MATEC Web Conf 119:1050

20. Liu W et al (2020) Lifecycle operational reliability assessment of water distribution networks based on the probability density evolution method. Probab Eng Mech 59:103037
21. Yue Q, Ang AHS (2016) Nonlinear response and reliability analysis of tunnels under strong earthquakes. Struct Infrastruct Eng 12(5):618–630
22. Zhou H, Li J (2017) Overall collapse and reliability analysis of RC structures under stochastic seismic excitations. CRC Press, pp 193–199
23. Papadopoulos V, Kalogeris I (2016) A Galerkin-based formulation of the probability density evolution method for general stochastic finite element systems. Comput Mech 57(5):701–716
24. Lucchesi M, Pintucchi B, Zani N (2019) The generalized density evolution equation for the dynamic analysis of slender masonry structures. Trans Tech Publ
25. Chen J, Li J (2005) Extreme value distribution and reliability of nonlinear stochastic structures. Earthq Eng Eng Vib 4(2):275–286
26. Li J, Chen JB (2005) Dynamic response and reliability analysis of structures with uncertain parameters. Int J Numer Meth Eng 62(2):289–315
27. Chen J et al (2007) Stochastic seismic response and reliability analysis of base-isolated structures. J Earthquake Eng 11(6):903–924
28. Fang X et al (2012) Probability density evolution method for stochastic earthquake response and reliability analysis of large-scale aqueduct structures. Appl Mech Mater 193–194:1230–1233
29. Fang X, Liu ZJ (2012) Stochastic response analysis and reliability evaluation of nonlinear structures under earthquake. Appl Mech Mater 166–169:2100–2104
30. Peng Y, Mei Z, Li J (2014) Stochastic seismic response analysis and reliability assessment of passively damped structures. J Vib Control 20(15):2352–2365
31. Liu Y, Yi H, Chen L (2014) Submarine pressure hull butt weld fatigue life reliability prediction method. Mar Struct 36:51–64
32. Ang AHS, Fan W (2017) Reliability-based maintenance of complex structures for life-cycle performance. CRC Press, pp 21–36
33. Guo Y-F et al (2018) The instability probability density evolution of the bistable system driven by Gaussian colored noise and white noise. Phys A 503:200–208
34. Li J (2009) Dynamic response and reliability analysis of wind-excited structures. Springer Netherlands, Dordrecht, pp 529–536
35. Li J, Chen H, Chen J (2006) Reliability analysis of prestressed egg-shaped digester. Springer Netherlands, Dordrecht, pp 422–422
36. Yang L et al (2017) Efficient probabilistic risk assessment for aeroengine turbine disks using probability density evolution. AIAA J 55(8):2755–2761
37. Tao W, Basu B, Li J (2018) Reliability analysis of active tendon-controlled wind turbines by a computationally efficient wavelet-based probability density evolution method. Struct Control Health Monitor 25(3):e2078-n/a
38. Nielsen SRK et al (2013) Failure analysis of wind turbines by probability density evolution method. Key Eng Mater 569–570:579–586
39. Devathi H, Sarkar S (2016) Study of a stall induced dynamical system under gust using the probability density evolution technique. Comput Struct 162:38–47
40. Zhang H, Xu Y (2018) Probabilistic load flow calculation by using probability density evolution method. Int J Electr Power Energy Syst 99:447–453
41. Su G et al (2015) A Gaussian process-based response surface method for structural reliability analysis. Struct Eng Mech 56:549–567
42. Yang D, Liu L (2014) Reliability analysis of structures with complex limit state functions using probability density evolution method. Struct Multidiscip Optim 50(2):275–286
43. Mei Z, Guo Z (2018) Verification of probability density evolution method through shaking table tests of a randomly base-driven structure. Adv Struct Eng 21(3):514–528
44. Kalogeris I, Papadopoulos V (2018) Limit analysis of stochastic structures in the framework of the probability density evolution method. Eng Struct 160:304–313
45. Afshari SS et al (2020) Time-varying structural reliability assessment method: application to fiber reinforced composites under repeated impact loading. Compos Struct, p 113287

46. Chen J-B, Li J (2009) A note on the principle of preservation of probability and probability density evolution equation. Probab Eng Mech 24(1):51–59
47. Li J, Chen J (2008) The principle of preservation of probability and the generalized density evolution equation. Struct Saf 30(1):65–77
48. Chen G, Yang D (2019) Direct probability integral method for stochastic response analysis of static and dynamic structural systems. Comput Methods Appl Mech Eng 357:112612
49. Lewis GN (1908) LIX. A revision of the fundamental laws of matter and energy. The London, Edinburgh, and Dublin. Phil Maga J Sci 16(95):705–717
50. Jordan R, Kinderlehrer D, Otto F (1998) The variational formulation of the Fokker-Planck equation. SIAM J Math Anal 29(1):1–17
51. Li J, Chen J (2009) Stochastic dynamics of structures. Wiley
52. Fakharian O, Salmani H, Kordkheili SAH (2019) A lumped parameter model for exponentially tapered piezoelectric beam in transverse vibration. J Mech Sci Technol 33:2043–2048. https://doi.org/10.1007/s12206-019-0407-x
53. Hosseini Kordkheili SA, Salmani H, Afshari SSG (2016) A stabilized piezolaminated nine-nodded shell element formulation for analyzing smart structures behaviors. Mech Adv Mater Struct 23(2):187–194
54. Nobahari H, Hosseini Kordkheili SA, Afshari SS (2014) Hardware-in-the-loop optimization of an active vibration controller in a flexible beam structure using evolutionary algorithms. J Intell Mater Syst Struct 25(10):1211–1223
55. Wang B et al (2018) A hybrid prognostics approach for estimating remaining useful life of rolling element bearings. IEEE Trans Reliab
56. Afshari SS, Liang X (2019) Machine learning based dynamic failure criteria for reliability analysis of bearings. In: 2019 prognostics and system health management conference (PHM-*Qingdao*)
57. Begelinger A, De Gee AWJ (1982) A study of the effect of radial clearance, contact angle and contact pressure on the wear of boundary-lubricated bearing bronze. Wear 77(1):45–56
58. Shah DS, Patel VN (2014) A review of dynamic modeling and fault identifications methods for rolling element bearing. Proc Technol 14:447–456

Rotating Machinery Condition Monitoring Methods for Applications with Different Kinds of Available Prior Knowledge

Stephan Schmidt and P. Stephan Heyns

Abstract Intelligent (or smart) condition monitoring methods make it possible to automatically infer the condition of the machine. However, the performance of the intelligent condition monitoring methods is much dependent on the available historical data. Different applications may have different levels of historical data available. Intelligent condition monitoring methods allow automatic detection and overcome the need for feature engineering. However, feature engineering is often misconstrued as being equivalent to engineering knowledge. In this chapter, it is proposed that the available engineering knowledge and intelligent condition monitoring methods need to be combined to obtain effective condition monitoring methods, i.e. where reliable fault detection, identification and trending can be performed. Engineering knowledge and historical data are referred to as prior knowledge in this work and methods are proposed to utilise it for fault diagnosis. An investigation is performed on gearbox data generated under time-varying operating conditions, with the importance of using prior knowledge highlighted.

Keywords Intelligent condition monitoring · Gearbox fault diagnosis

1 Introduction

The development of intelligent condition monitoring methods is important for industries where the safety and reliability of expensive equipment are important (e.g. wind turbine gearboxes, steam turbines). Intelligent condition monitoring methods make it possible to automatically infer the condition of the machine by using the available historical data for training (or optimising) the models.

S. Schmidt (✉) · P. S. Heyns
Centre for Asset Integrity Management, University of Pretoria, Pretoria, South Africa
e-mail: stephan.schmidt@up.ac.za

P. S. Heyns
e-mail: stephan.heyns@up.ac.za

F. Chaari et al. (eds.), *Smart Monitoring of Rotating Machinery for Industry 4.0*, Applied Condition Monitoring 19, https://doi.org/10.1007/978-3-030-79519-1_6

Intelligent condition monitoring methods typically make use of state-of-the-art machine learning and deep learning models developed in the computer science field [6, 19]. Unsupervised learning methods can be used to infer the salient information in the data without supplying condition labels, while supervised learning methods can be used for automatic condition inference (e.g. what is the condition of the machine) [9]. However, many computer vision problems can be approximated as closed set recognition or classification problems, while condition monitoring problems are inherently open set in most practical applications (especially for expensive machines) [12]. An open set condition recognition problem refers to the case where historical data of only some of the damage modes that can be encountered are available. Therefore, care should be used when applying closed set recognition algorithms in the condition monitoring field [12].

Even though the intelligent condition monitoring methods aim to overcome the need for engineering knowledge, much research has been conducted to understand the statistical characteristics of vibration signals (i.e. cyclostationary) and how fault information manifest in vibration signals (e.g. changes in instantaneous power) [1, 2]. In discrepancy analysis, data-driven models of a healthy machine are combined with signal analysis methods that exploit the knowledge about the nature of fault signals (e.g. faults are cyclostationary with a known period) for more effective novelty detection [5]. Utilising the available engineering knowledge becomes especially important when the historical data are scarce. Hence, in this work, we identify different levels of prior knowledge and suggest methods to utilise the available prior knowledge for performing effective condition monitoring.

The layout of the chapter is as follows: In Sect. 2, the availability of different kinds of prior knowledge is discussed, whereafter an investigation is performed in Sect. 3 to illustrate how prior knowledge can be used for effective condition monitoring. Finally, the work is concluded and recommendations are made in Sect. 4.

2 Prior Knowledge in Condition Monitoring

Many condition monitoring methods have been developed over the past few decades and the methods typically range from purely engineering knowledge-based methods (e.g. signal processing) to deep learning methods, capable of automatically identifying and extracting the salient information from the data [3, 6, 9]. The applicability and the performance of the methods inherently depend on the available prior knowledge. Prior knowledge in this context refers to information that is available before the condition is inferred or before the data are analysed.

Much research has been performed on understanding the statistical properties of the vibration signals acquired from common rotating machine components (e.g. bearings and gears). However, this engineering prior knowledge (e.g. the statistical properties of the signal, bearing fault frequencies, drive-train layout) can be difficult to utilise for automatic fault diagnosis. In contrast, classification-based data-driven methods (e.g. statistical learning to deep learning methods) can perform automatic condition inference, but historical data from a healthy machine and different failure modes are needed a priori. This can result in a class imbalance problem and an open set recognition problem.

In this work, we suggest that all of the available prior knowledge (i.e. engineering knowledge and historical data) need to be utilised to ensure that effective condition monitoring (i.e. reliable fault detection, fault identification and fault trending) can be performed. This is illustrated in Fig. 1.

We divide the rest of this section in terms of engineering knowledge and knowledge that can be extracted with machine learning methods.

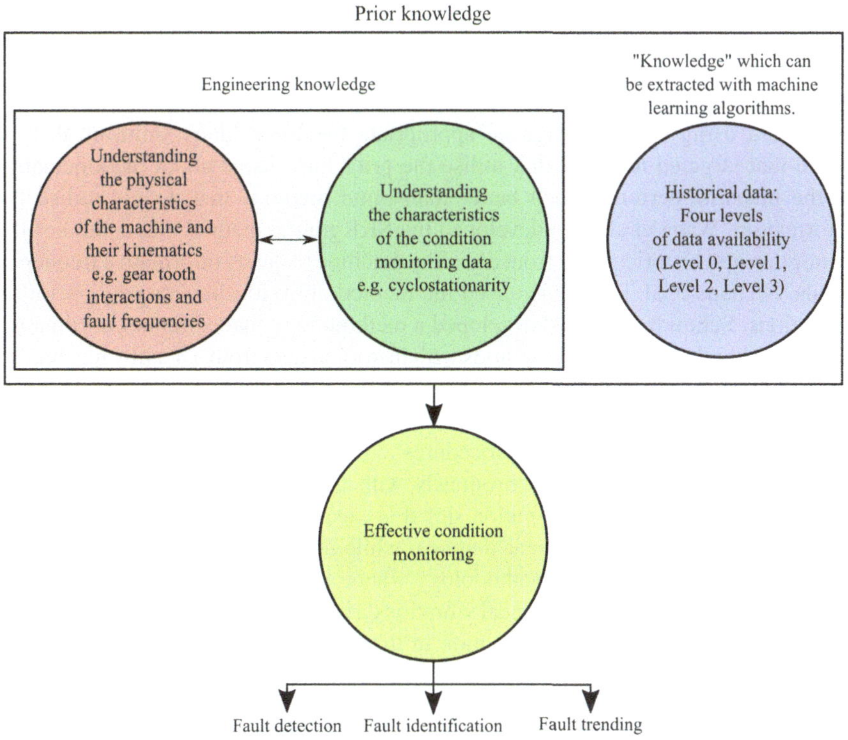

Fig. 1 Effective condition monitoring by utilising the available prior knowledge

2.1 Engineering Knowledge

Vibration signals acquired from rotating machines are typically angle cyclostationary [2] and the impulses become angle-time cyclostationary under time-varying operating conditions [1]. This makes it possible to utilise advanced signal analysis techniques such as the spectral coherence, the squared envelope spectrum and the instantaneous power spectrum for extracting or highlighting fault information in the vibration data. Antoni and Borghesani [3] developed a statistical methodology to design condition indicators by performing hypothesis tests under different assumptions of the statistical nature of the vibration signals. Gryllias and Antoniadis [4] used a physics-based model (i.e. engineering knowledge) to generate data which were subsequently used to train support vector machines to infer the condition of failures not observed in the past.

It is possible to enhance the fault information by identifying informative frequency bands and using this to design an appropriate bandpass filter. Smith et al. [17] found that targeted methods that utilise the prior knowledge about the kinematics of the machine perform much better than blind methods that do not utilise the information. Wang et al. [18] developed the SKRgram,a method that combines the kurtogram and historical data from a healthy machine to detect informative frequency bands. Niehaus et al. [11] generalised the methodology for time-varying operating conditions. Schmidt et al. [14] developed a methodology that combines informative frequency band identification methods with historical data from a healthy machine to enhance the fault information. By combining the engineering knowledge (i.e. faults manifest as impulsive components in narrow frequency bands) with historical data, it is possible to find more robust procedures to enhance the fault information.

Gear damage manifests synchronously with the connected rotating shaft. The synchronous average of the vibration signal is capable of highlighting synchronous changes in the signal due to gear damage, while attenuating the non-synchronous components. [5], proposed a methodology where a discrepancy signal is generated from a model of the healthy historical vibration data. The discrepancy signal measures the time-localised anomaly information in the signal. Thereafter, the synchronous average of the discrepancy signal is calculated to visualise the anomalous components. This representation is very effective to visualise the condition of the gears in the gearbox. Therefore, the synchronous average, which is common in signal processing-based condition monitoring, was combined with a data-driven model for more effective condition monitoring. Schmidt et al. [15] extended this discrepancy analysis approach for bearings,where the spectrum of the discrepancy signal highlighted the periodicity of the anomalous components and therefore can be used to identify the component that is damaged. Hence, the signal analysis methods (e.g. synchronous averaging) makes it possible to interpret the outputs from discrepancy analysis-based data-driven models for more effective condition monitoring.

Under time-varying speed conditions, the frequency and amplitude modulation impede the application of conventional methods. It is therefore desirable to analyse the signal in the angle domain and to attenuate the amplitude modulation due to varying operating conditions. In some applications, the rotational speed is not available and therefore the speed need to be estimated from the vibration signal. Leclere et al. [8] developed a multi-order probabilistic approach for speed estimation that utilises prior knowledge about the kinematics of the machine as well as prior knowledge about the operating condition range of the machine. The authors found that by utilising this prior knowledge, more reliable speed estimates can be obtained.

However, to ensure that effective condition monitoring can be performed, it is important to combine this engineering knowledge with the available historical data.

2.2 *Knowledge Extracted from Machine Learning Algorithms*

The different kinds of historical data availability are illustrated in Fig. 2. The population of damage modes refers to all possible damage modes that can be encountered for the machine and the different levels of available historical data are regarded as prior knowledge.

It is important to be cognizant about the difference between common computer vision problems and the condition monitoring problem; in condition monitoring, there is a continuous transition within a specific damage mode from an approximately

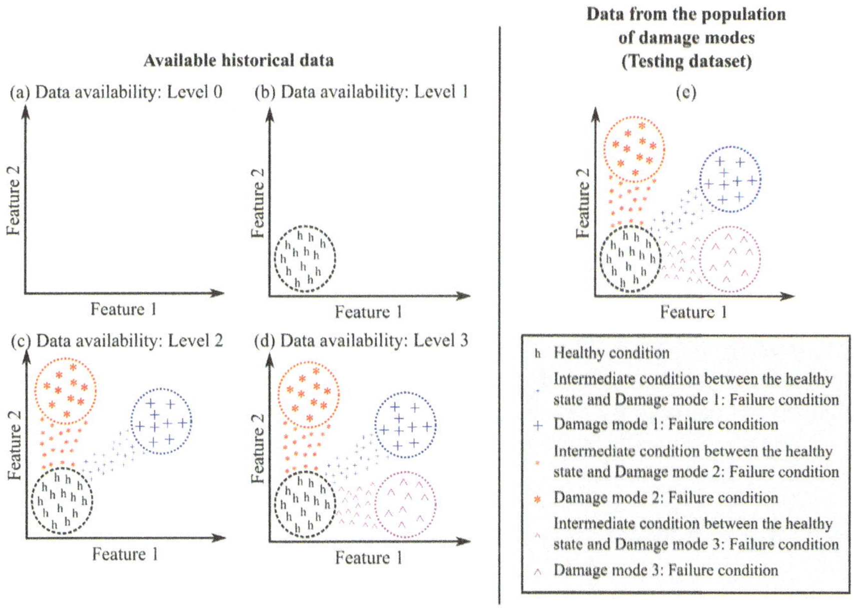

Fig. 2 Different Data Availability Levels (DAL) that can be encountered in condition monitoring illustrated with a two-dimensional feature space

healthy condition to a failure state. In computer vision, there are usually discrete states (e.g. a number that needs to be identified, the animal) without any transitions between the discrete classes and therefore the conventional classification algorithms may not address the underlying physics that govern the transition in the data [12].

Different models can be used to extract information from the data and/or can be used to model the data (e.g. statistical models to multi-layered neural networks), with only a short discussion given here. Often in condition monitoring, statistical models such as kernel density estimators, Gaussian mixture models and hidden Markov models are used to model processed data [5, 16]. Multi-layered neural networks can automatically extract the salient information from the data and therefore raw data are often supplied to the models. This reduces the need for performing manual feature extraction and feature selection. Some recent discussions on machine learning and deep learning for condition monitoring applications are given by Zhao et al. [19] and Lei et al. [10].

In the next subsections, we provide an overview of the problem encountered in different Data Availability Levels (DALs). Even though there are many potential methods to combine different kinds of prior knowledge for effective condition monitoring (e.g. [5, 14, 18]), a simple method is proposed in this work to illustrate the important concepts.

2.2.1 Data Availability: Level (DAL) 0

In DAL 0, there are no historical data to train a data-driven method. And even though unsupervised learning methods may be used to extract the underlying structure of the data, engineering knowledge is required to interpret the results. Schmidt and Heyns [13] proposed a divergence analysis method that can be used for automated localised gear damage detection by utilising the available prior knowledge about the statistical characteristics of vibration signals generated by localised faults.

In this work, we first calculate the standardised Squared Envelope Spectrum (SES) with the procedure used by Kass et al. [7]. Thereafter, we apply a method to automatically select the threshold for detection. By assuming that the kinematics of the gearbox and rotational speed are available a priori, we monitor specific components (e.g. bearing defect frequencies, gear defect frequencies). If an amplitude exceeds the threshold, it indicates that there is a strong periodicity in the data, which could be indicative of damage. By monitoring the signal component over time, it is possible to determine whether the mechanical component is deteriorating.

2.2.2 Data Availability: Level (DAL) 1

In DAL1, it is possible to utilise the healthy historical data to train a model to perform automatic novelty detection, but also to identify the frequency of the novelty components with discrepancy analysis [5]. Another method is to enhance the novel frequency bands for fault diagnosis [14] which results in an enhanced vibration signal. Schmidt et al. [16] developed a method that combines the spectral coherence, historical data from a healthy machine, and knowledge about the fault frequencies of the important mechanical components for novelty detection under time-varying operating conditions. By combining engineering knowledge with data-driven models, it is possible to perform automatic fault diagnosis.

In this work, we extended the DAL 0 procedure to include historical data from a healthy machine as follows:

1. Calculate the SES and standardise it as performed in DAL 0.
2. Extract the harmonics of the critical mechanical components as performed in DAL 0.
3. Train a Gaussian model on the harmonics extracted from a healthy machine.
4. Calculate the Mahalanobis distance for the harmonics extracted from the new data. The Mahalanobis distance is a scalar quantity that captures the condition of the gearbox and it allows for more effective monitoring.

2.2.3 Data Availability: Level (DAL) 2

In DAL 2, it is assumed that healthy historical data are available as well as the failure data associated with some of the damage modes that can be encountered in the machine. Since the machine condition monitoring problem is an open set recognition problem, with continuous transitions from a healthy state to a failure state, careful consideration needs to be given to the model. In Schmidt and Heyns [12] it was indicated that the data follows a stochastic transition during the condition degradation process. Gaussian Mixture Models (GMMs), hidden Markov models and mixture density networks are capable of capturing the expected distributions in the data.

In this work, we propose the following procedure:

1. Follow the procedure in DAL 1 to fit the model associated with the healthy data.
2. Extract the amplitudes of the fundamental components of the mechanical components-of-interest from the historical fault data.
3. For each available historical fault data case (e.g. outer race bearing damage, inner race bearing damage), train a separate GMM on the datasets. The GMM therefore captures the whole transition of the machine and the latent states of the GMM capture the underlying state of the machine (e.g. incipient damage, intermediate damage, severe damage).
4. Use the posterior probability of the condition

$$P(C_i|x) = \frac{p(x|C_i)P(C_i)}{p(x)} \tag{1}$$

where C_i denotes the ith class and x denotes the features, to infer the condition of the machine. However, if the likelihood of the models is too small, then the class label is rejected and a novelty is detected. Hence, the label (or condition) is only assigned to the machine if there is strong evidence that this is true. In this work, the prior probability $P(C_i)$ of the healthy class is made five times larger than the healthy machine. This also reduces the probability of false alarms. A more detailed discussion of the open set recognition problem is given by Schmidt and Heyns [12].

2.2.4 Data Availability: Level (DAL) 3

In DAL 3, it is assumed that the historical data from all damage modes are available for determining the condition of the machine. This reduces the condition recognition problem, to a condition classification problem (i.e. it is only necessary to determine the most probable class). Many methods are proposed in the literature that directly addresses this problem. However, many of these approaches assign class labels to specific machine conditions (e.g. an outer race bearing defect of 0.007 inches is assigned a different class label than an outer race bearing defect of 0.014 inches). This raises two questions:

1. How can these class labels be obtained on industrial machines where the defect size typically cannot be measured during the operation of the machine?
2. What is the class label if the machine condition is between two classes (e.g. if the actual defect size is 0.010 inches in the previous example)? Often, the deep learning methods are trained to optimally separate the classes (or conditions) available during training. This optimal separation is usually performed with a non-linear mapping to a new latent space. It is unreasonable to assume that the data would lie exactly between the two classes in the feature space if a general non-linear mapping is performed by deep learning methods.

Instead of using the conventional approaches, it proposed that in a DAL 3 scenario, the proposed DAL 2 approach is used as well. This ensures that the class label can be consistently assigned to the machine as it deteriorates and it is more effective since it utilises the models trained in DAL 2.

3 Case Study

The phenomenological gearbox data investigated in the paper by Schmidt et al. [14] are considered in this work for a gearbox where it is possible to have distributed gear damage, inner race bearing damage and outer race bearing damage. We assume that the kinematics of the gearbox is known and accurate rotational speed measurements are available. Of course, it is possible to use the multi-order probabilistic approach [8] if rotational speed measurements are not available. All measurements were generated by simulating a ramp-up scenario as follows:

$$w = 5t + 10 \tag{2}$$

where w is the speed of the shaft in Hz and $0 \leq t < 5$. The purpose of this investigation is to illustrate some important concepts related to the different data availability levels.

3.1 Data Availability: Level 0

In this section, the method proposed in Sect. 2 is used to infer the condition of the numerical gearbox. The gear, which is connected to the reference shaft, has distributed damage, which would manifest as a random component in the signal at one shaft order and its harmonics. Firstly, the standardised SES of the order tracked vibration signal is calculated, whereafter a threshold is determined for automatic fault detection. The percentage of Points Exceeding the Threshold (PET) in the standardised SES is compared against the threshold in Fig. 3a with the selected threshold being shown. The resulting threshold is superimposed on the standardised SES in Fig. 3b, with the signal components associated with the fundamental Shaft Order (SO), Ball-Pass Order of the Outer race (BPOO) and the Ball-Pass Order of the Inner race (BPOI) shown as well.

It is evident from the results in Fig. 3b that only the distributed gear component exceeds the threshold. This procedure is performed for 150 measurements, with the gear deteriorating over time with the result presented in Fig. 3c. If the different signal components are compared against the threshold, it is seen in Fig. 3d that the gear exceeds the threshold, with one false alarm in the BPOI. This false alarm is easily removed if the alarm is only triggered when several consecutive measurements

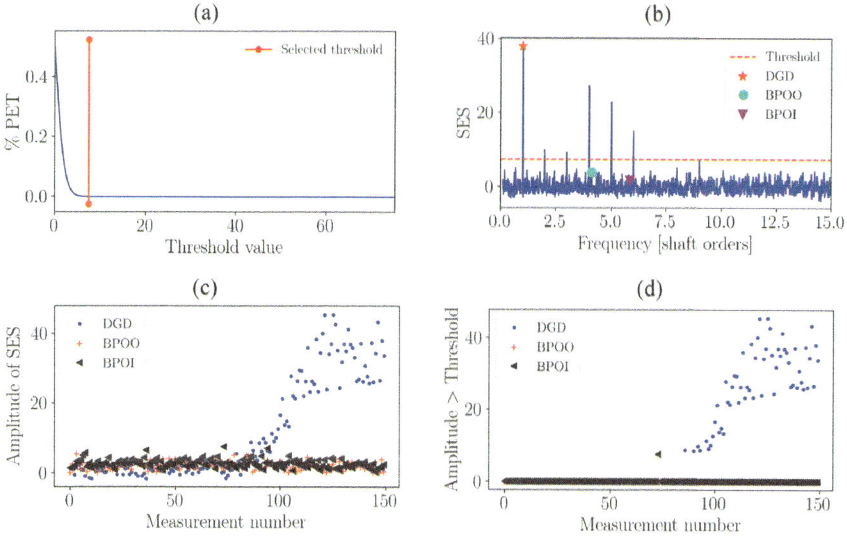

Fig. 3 An illustration of a DAL 0 approach for fault diagnosis

exceed the threshold. Hence, when comparing the raw amplitude (Fig. 3c) to the processed amplitude (Fig. 3d), we see that it is possible to automatically monitor specific components in the gearbox with this approach.

3.2 Data Availability: Level 1

In the DAL 1 investigation, only the BPOO and the DGD are monitored to ensure that the concept can be easily visualised. In this case, it is assumed that measurements of the healthy gearbox are available, whereafter the gear deteriorated and the bearing developed outer race bearing damage in two separate cases. The resulting features of the healthy and the damaged gearboxes are shown in Fig. 4a and b. The DGD case contains the same data as the previous investigation.

The stochastic transition from the healthy state to the failure states, discussed in the paper by Schmidt and Heyns [12], is evident for the two cases. The Mahalanobis distance is calculated over the measurement number for the gear and the bearing, with the results presented in Fig. 4a and b. The healthy gearbox data were used to select the appropriate threshold for detection. The results indicate that it is possible to only monitor the Mahalanobis distance, and if anomalous behaviour is detected, it is possible to investigate the different features to determine the condition of the gearbox.

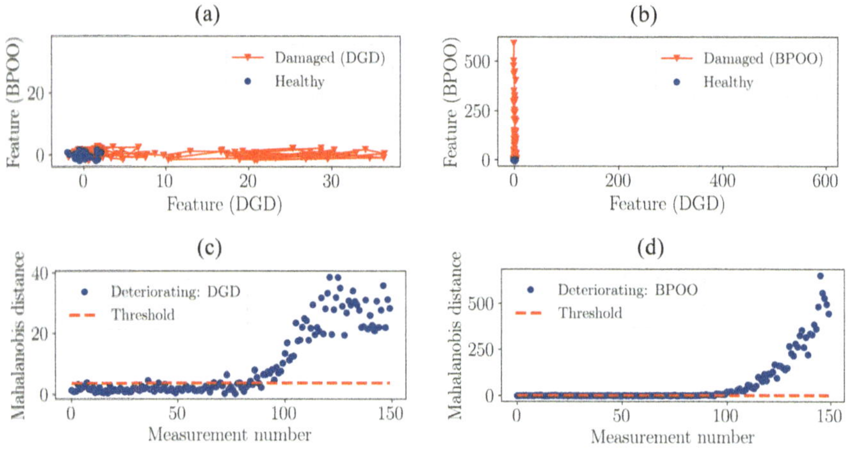

Fig. 4 The features acquired from a deteriorating gear and bearing and the corresponding Mahalanobis distances are presented

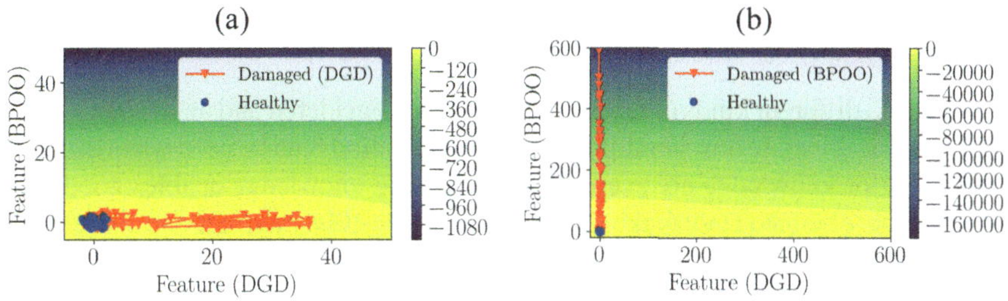

Fig. 5 The log-likelihood of the distributed gear model for the case where the gearbox had distributed gear damage **(a)** and a damaged bearing **(b)**

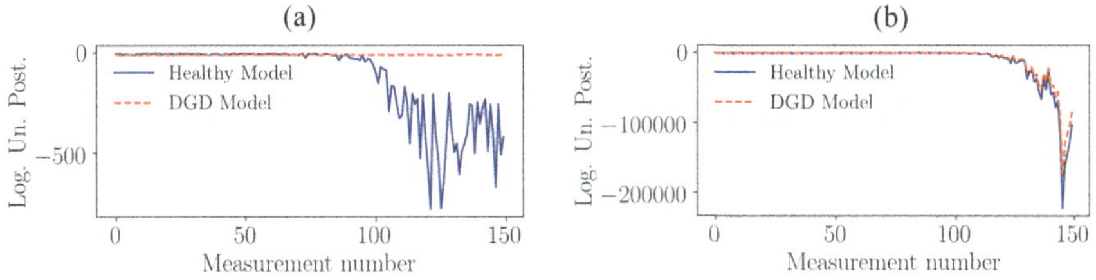

Fig. 6 The logarithm of the unnormalised posterior distribution, calculated with the numerator of Eq. (1), are presented for a gearbox with distributed gear damage **(a)** and a damaged bearing **(b)**

3.3 Data Availability: Level 2

In the DAL 2 investigation, it is assumed that healthy historical data and historical fault data, acquired from a gearbox that suffered from a damaged gear, is available. The log-likelihood of the GMM of the damaged gear is presented in Fig. 5a and b.

Since the bearing damage does not form part of the historical dataset, the GMM of the damaged gear has very small likelihood values in Fig. 5b.

In Fig. 6, the logarithm of the unnormalised posterior probability is presented (i.e. the numerator of Eq. 1) for the healthy model and the model of the damaged gear for the deteriorating gear and bearing. In Fig. 6a, the gear deteriorated over measurement number, which is the reason why the DGD model has a much larger log unnormalised probability. In Fig. 6b, the bearing suffered from outer race damage. Since historical data of this damage mode is not available in the training dataset, both the healthy model and the distributed gear models are incapable of inferring the condition and therefore very small values are obtained by both models. This means that the predicted class (distributed gear damage) should be rejected, which means that the data comes from a new class, i.e. the gearbox is not healthy and does not have distributed gear damage. By using the DAL 2 approach, it is therefore possible to recognise that the gearbox is unhealthy and to recognise that a new machine condition is encountered.

4 Conclusions and Recommendations

In this work, different kinds of prior knowledge are considered and an investigation is performed to illustrate how condition monitoring methods can be used when different levels of prior knowledge are available. Engineering knowledge enables good indicators to be obtained and makes it possible to sensibly interpret the data, while data-driven methods make it possible to extract and learn the salient information in the data for making predictions. The literature study and the results indicate that by combining engineering prior knowledge (e.g. cyclostationarity, order tracking) with historical data (which can be utilised with intelligent algorithms), more effective condition monitoring can be performed.

It is recommended that different methods for performing open set recognition need to be investigated and compared in future work. Ultimately, the performance of the methods depends on the extracted or learned features and the ability to capture the underlying densities of the known conditions. Deep learning models could improve our ability to model the densities of rotating machine data. The performance of the methods significantly depends on the threshold selection procedure and therefore different threshold selection procedures must be investigated and compared for condition monitoring applications. Model re-calibration methods (e.g. if maintenance was performed on the machine) need to be investigated as well.

Acknowledgements The South African authors gratefully acknowledge the support that was received from the Eskom Power Plant Engineering Institute (EPPEI) in the execution of this research.

References

1. Abboud D, Baudin S, Antoni J, Rémond D, Eltabach M, Sauvage O (2016) The spectral analysis of cyclo-non-stationary signals. Mech Syst Signal Process 75:280–300
2. Antoni J (2009) Cyclostationarity by examples. Mech Syst Signal Process 23(4):987–1036
3. Antoni J, Borghesani P (2019) A statistical methodology for the design of condition indicators. Mech Syst Signal Process 114:290–327
4. Gryllias KC, Antoniadis IA (2012) A Support vector machine approach based on physical model training for rolling element bearing fault detection in industrial environments. Eng Appl Artif Intell 25(2):326–344
5. Heyns T, Heyns PS, De Villiers JP (2012) Combining synchronous averaging with a Gaussian mixture model novelty detection scheme for vibration-based condition monitoring of a gearbox. Mech Syst Signal Process 32:200–215
6. Jia F, Lei Y, Lin J, Zhou X, Lu N (2016) Deep neural networks: a promising tool for fault characteristic mining and intelligent diagnosis of rotating machinery with massive data. Mech Syst Signal Process 72:303–315
7. Kass S, Raad A, Antoni J (2019) Self-running bearing diagnosis based on scalar indicator using fast order frequency spectral coherence. Meas 138:467–484
8. Leclère Q, André H, Antoni J (2016) A multi-order probabilistic approach for Instantaneous Angular Speed tracking debriefing of the CMMNO'14 diagnosis contest. Mech Syst Signal Process 81:375–386

9. Lei Y, Jia F, Lin J, Xing S, Ding SX (2016) An intelligent fault diagnosis method using unsupervised feature learning towards mechanical big data. IEEE Trans Industr Electron 63(5):3137–3147
10. Lei Y, Yang B, Jiang X, Jia F, Li N, Nandi AK (2020) Applications of machine learning to machine fault diagnosis: a review and roadmap. Mech Syst Signal Process 138:106587
11. Niehaus WN, Schmidt S, Heyns PS (2020) NIC Methodology: a probabilistic methodology for improved informative frequency band identification by utilizing the available healthy historical data under time-varying operating conditions. J Sound Vib 488:115642
12. Schmidt S, Heyns PS (2019) An open set recognition methodology utilising discrepancy analysis for gear diagnostics under varying operating conditions. Mech Syst Signal Process 119:1–22
13. Schmidt S, Heyns PS (2019) Localised gear anomaly detection without historical data for reference density estimation. Mech Syst Signal Process 121:615–635
14. Schmidt S, Heyns PS, Gryllias KC (2019) A pre-processing methodology to enhance novel information for rotating machine diagnostics. Mech Syst Signal Process 124:541–561
15. Schmidt S, Heyns PS, Gryllias KC (2019) A discrepancy analysis methodology for rolling element bearing diagnostics under variable speed conditions. Mech Syst Signal Process 116:40–61
16. Schmidt S, Heyns PS, Gryllias KC (2020) A methodology using the spectral coherence and healthy historical data to perform gearbox fault diagnosis under varying operating conditions. Appl Acoust 158:107038
17. Smith WA, Borghesani P, Ni Q, Wang K, Peng Z (2019) Optimal demodulation-band selection for envelope-based diagnostics: a comparative study of traditional and novel tools. Mech Syst Signal Process 134:106303
18. Wang T, Han Q, Chu F, Feng Z (2016) A new SKRgram based demodulation technique for planet bearing fault detection. J Sound Vib 385:330–349
19. Zhao R, Yan R, Chen Z, Mao K, Wang P, Gao RX (2019) Deep learning and its applications to machine health monitoring. Mech Syst Signal Process 115:213–237

Model Based Fault Diagnosis in Bevel Gearbox

Palash Dewangan, Dada Saheb Ramteke, and Anand Parey

Abstract Bevel gearboxes are used in many industrial, automotive, and aerospace applications for transmission purposes. A gearbox should operate without any malfunction to achieve smooth and high performance. Any incipient fault in a gearbox may grow severe, which may lead to high noise and vibration of the gearbox and subsequently lead to failure of the gearbox. It is, therefore, essential to detect the incipient faults at the earliest to avoid premature failures. The effect of the tooth fault is reflected in gear mesh stiffness. In this paper, a mesh stiffness model of missing tooth fault in a bevel gear is proposed. The dynamic response of one stage bevel gearbox with a missing tooth fault is computed to identify the fault feature characteristics of the bevel gearbox. The simulation results show some distinct time domain and frequency domain characteristics for the identification of faults. The simulation results are compared with the vibration responses obtained from the experiment in both the time and frequency domain. The comparison of experimental and simulation results show that the proposed model successfully identifies the missing tooth fault in a bevel gearbox.

Keywords Bevel gear modeling · Missing tooth fault · Time-varying mesh stiffness · Dynamic response

1 Introduction

Modeling of faults in a gearbox is a promising way to understand the dynamic response characteristics of a gearbox under malfunctioning [1]. Tooth breakage is a tooth surface failure of gears. Partial tooth breakage (chipping) or complete tooth

P. Dewangan · D. S. Ramteke · A. Parey (✉)
Department of Mechanical Engineering, Indian Institute of Technology Indore, Indore, India
e-mail: anandp@iiti.ac.in

P. Dewangan
e-mail: phd1701103004@iiti.ac.in

D. S. Ramteke
e-mail: phd1601103002@iiti.ac.in

F. Chaari et al. (eds.), *Smart Monitoring of Rotating Machinery for Industry 4.0*,
Applied Condition Monitoring 19, https://doi.org/10.1007/978-3-030-79519-1_7

breakage (missing tooth) occurs when sudden extremely large stresses develop at the tooth surfaces [2]. Tooth breakage may also occur due to high tooth impact loads [3]. Gear mesh stiffness is the primary source of excitation in the gear transmissions and, consequently, is the reason for noise and vibrations. In addition, the presence of any fault in gears, which is reflected as the change in the gear mesh stiffness, makes the dynamic behavior of the gears more complex. Thus, it is imperative to understand the dynamic behavior of the gear systems under the presence of faults.

In the last two decades, a few works [4–9] have been published on the modeling of bevel gears without considering faults. These works are mainly focused on the modeling of teeth deformation due to contact, static, and dynamic analysis and effects of assembly and manufacturing error on the dynamic response. A comprehensive review of the modeling of gear faults is presented by Liang and co-authors [10]. However, from their review paper, it was revealed that there is a scarcity in the literature on the modeling of faults in bevel gears, especially straight bevel gears. Therefore, literature considering the modeling of faults in bevel gears is summarized here. Karray and co-authors [11] presented a dynamic model of one-stage spiral bevel gear for calculating dynamic response with tooth crack defect. Yassine and co-authors [12] developed a three dimensional dynamic model of a two-stage straight bevel gearbox with manufacturing and tooth crack defect. Recently, Karay and co-authors [13] performed a dynamic response analysis of a spiral bevel gear system under nonstationary operations in the presence of a tooth crack defect. In a recent paper by Ramteke and co-authors [14] worked on the identification of micron-level wear in bevel gears.

In addition, tooth surface failure such as wear and tooth breakage occurs under high dynamic loads and uncertain loading conditions [15–17]. Tooth breakage (chipping and missing of the tooth) is a severe damage condition and may lead to rapid failure of the gearbox. In the available literature, modeling of missing tooth fault and its effects on the dynamics of bevel gears has not been addressed. This motivates the authors to study the dynamic behavior of bevel gearbox with missing tooth fault. However, modeling of missing tooth faults in spur gears can be found in the work of Tian and co-authors [3]. They proposed that the teeth will have only a single tooth pair contact (STPC) instead of double tooth pair contact (DTPC), and there will be no contact instead of STPC. However, the results were not validated through the experiments. More recently, a few researchers [18–20] have presented a mesh stiffness model for missing tooth fault for spur and planetary gears using potential energy method. In these papers, the time varying mesh stiffness obtained is similar to the Tian's (2004) model.

In this study, a time-varying mesh stiffness (TVMS) model of straight bevel gears to account for a missing tooth fault is proposed based on the approach of Tian and co-authors [3] for simplicity. The TVMS model is incorporated in the dynamic model of a one-stage bevel gearbox to calculate dynamic responses. Experiments are conducted to validate the simulation results. It is shown that the model is capable of diagnosing the missing tooth fault in a one-stage straight bevel gearbox.

2 Dynamic Modelling of One Stage Straight Bevel Gearbox

In this study, a dynamic model of a straight bevel gearbox is adopted from [12] as shown in Fig. 1. The modeling of the gearbox has the following assumptions. The stiffness due to meshing of the gears is represented as the linear springs and mesh damping is neglected. Effect of friction force during meshing is ignored. Any assembly and manufacturing errors are ignored. Transmission error is not considered. The model consists of a bevel gear pair with each gear (pinion/wheel) having five degrees of freedom, i.e., three translations and two rotations. The translational displacements are defined by $x_j, y_j, z_j, (j = p, w)$ and angular displacements are defined as ϕ_j and ψ_j. The pinion is connected to the motor for input, and the wheel is connected to the receiver (load). The rotations of the pinion shaft and wheel shaft are defined as $\theta_j(j = p, w)$. The rotations of motor and receiver shafts are defined as θ_m and θ_r respectively. $k(t)$ is the time-varying mesh stiffness, k_{jx} and $k_{jy}(j = p, w)$ are the radial bearing stiffness, k_{jz} is the axial bearing stiffness, $k_{j\phi}$ and $k_{j\psi}$ are the bearing tilt stiffness, and $k_{j\theta}$ is the torsional stiffness of the shaft containing gears j.

The tooth deflection during meshing is given by [12]

$$\delta = \{L\}^T\{q\} \tag{1}$$

where

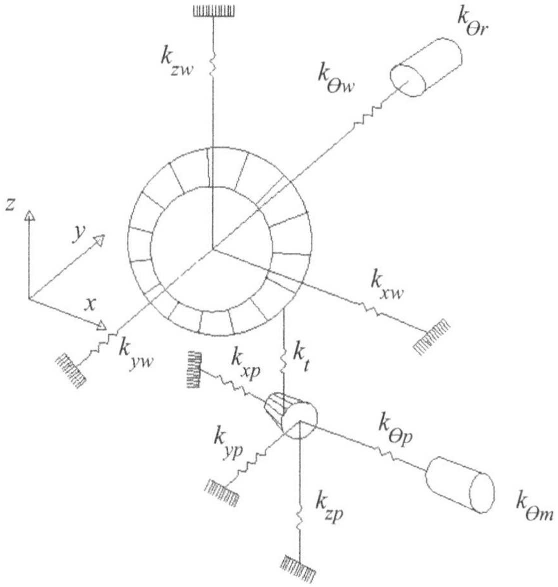

Fig. 1 A dynamic model of a one-stage bevel gear system

Table 1 Tooth deflection coefficients [12]

$c_1 = b_1 \sin(a_1 u_1)$
$c_2 = a_1 \cos(u_1)\sin(a_1 u_1) - \sin(u_1)\cos(a_1 u_1)$
$c_3 = a_1 \sin(u_1)\sin(a_1 u_1) + \cos(u_1)\cos(a_1 u_1)$
$c_4 = b_2 \sin(a_2 u_2)$
$c_5 = a_2 \cos(u_2)\sin(a_2 u_2) - \sin(u_2)\cos(a_2 u_2)$
$c_6 = a_2 \sin(u_2)\sin(a_2 u_2) + \cos(u_2)\cos(a_2 u_2)$
$c_7 = c_2 v b_1 \cos(a_1 u_1) - c_1 v[a_1 \cos(u_1)\cos(a_1 u_1) + \sin(u_1)\sin(a_1 u_1)]$
$c_8 = c_1 v[a_1 \sin(u_1)\cos(a_1 u_1) - \sin(u_1)\sin(a_1 u_1)] - c_3 v b_1 \cos(a_1 u_1)$
$c_9 =$ $c_3 v[a_1 \cos(u_1)\cos(a_1 u_1) + \sin(u_1)\sin(a_1 u_1)] - c_2 v[a_1 \sin(u_1)\cos(a_1 u_1) + \cos(u_1)\sin(a_1 u_1)]$
$c_{10} = c_4 v[a_2 \cos(u_2)\cos(a_2 u_2) + \sin(u_2)\sin(a_2 u_2)] - c_5 v b_2 \cos(a_2 u_2)$
$c_{11} = c_4 v[a_2 \sin(u_2)\cos(a_2 u_2) - \cos(u_2)\sin(a_2 u_2)] - c_6 v b_2 \cos(a_2 u_2)$
$c_{12} =$ $-c_6 v[a_2 \cos(u_2)\cos(a_2 u_2) + \sin(u_2)\sin(a_2 u_2)] + c_5[a_2 \sin(u_2)\cos(a_2 u_2) - \cos(u_2)\sin(a_2 u_2)]$

$$\{L\} = \{c_1, c_2, c_3, c_4, c_5, c_6, c_7, c_8, c_{10}, c_{11}, 0, c_9, c_{12}, 0\} \tag{2}$$

and

$$q = \left\{x_p, y_p, z_p, x_w, y_w, z_w, \phi_p, \psi_p, \phi_w, \psi_w, \theta_m, \theta_p, \theta_w, \theta_r\right\}^T \tag{3}$$

The coefficients $c_j (j = 1, 2, ...12)$ are given in Table 1. q is the generalized displacement vector, where subscripts p, w, m and r refers to the pinion, wheel, motor, and receiver respectively.

In Table 1, v_i is the radius of the spherical circle of the bevel gear geometry and u_i is parameter of the lead line, where $i = 1, 2,\ \ 1 - pinion,\ and\, 2 - wheel$, a_i and b_i are the parameter of the bevel gear geometry and can be defined as [4]

$$a_i = \sin(\delta_{bi}), \text{ and } b_i = \cos(\delta_{bi}) \tag{4}$$

where

(δ_{bi}) is the half top base cone angle.

The equation of motion of the system is obtained by Lagrange's Method and is written as [12]

$$M\ddot{q}(t) + C\dot{q}(t) + [K_s + K(t)]q(t) = F(t) \tag{5}$$

In Eq. (4), M is the mass matrix of the system and can be written as

$$M = diag\left(m_p, m_p, m_p, m_w, m_w, m_w, I_{px}, I_{py}, I_{wx}, I_{wy}, I_m, I_{p\theta}, I_{w\theta}, I_r\right) \tag{6}$$

where m_j $(j = p, w)$ is the lumped mass and I_{ij} $(i = x, y; j = p, w)$ is the mass moment of inertia of the gear bodies. I_m, $I_{p\theta}$, $I_{w\theta}$, and I_r are the mass moment of inertia of motor, pinion shaft, wheel shaft, and the load respectively.

The time-varying mesh stiffness matrix is given by [12]

$$K(t) = k(t)\{L\}\{L\}^T \tag{7}$$

The bearing stiffness matrix is given by [12]

$$K_s = \begin{bmatrix} K_T & 0 \\ 0 & K_R \end{bmatrix} \tag{8}$$

where

$$K_T = diag(\{k_{px}, k_{py}, k_{pz}, k_{wx}, k_{wy}, k_{wz}\}) \tag{9}$$

$$K_R = \begin{bmatrix} k_{p\phi} & 0 & 0 & 0 & 0 & 0 & 0 & 0 \\ 0 & k_{p\psi} & 0 & 0 & 0 & 0 & 0 & 0 \\ 0 & 0 & k_{w\phi} & 0 & 0 & 0 & 0 & 0 \\ 0 & 0 & 0 & k_{w\psi} & 0 & 0 & 0 & 0 \\ 0 & 0 & 0 & 0 & k_{p\theta} & -k_{p\theta} & 0 & 0 \\ 0 & 0 & 0 & 0 & -k_{p\theta} & k_{p\theta} & 0 & 0 \\ 0 & 0 & 0 & 0 & 0 & 0 & k_{w\theta} & -k_{w\theta} \\ 0 & 0 & 0 & 0 & 0 & 0 & -k_{w\theta} & k_{w\theta} \end{bmatrix} \tag{10}$$

The proportional Rayleigh damping C can be defined as

$$C = 0.05M + 10^{-4}K_m \tag{11}$$

where K_m is the mean mesh stiffness matrix.

3 Modelling of Mesh Stiffness Function

3.1 Mesh Stiffness Model of a Healthy Bevel Gear

In this paper, TVMS of a bevel gear pair is modeled as the square wave approximation. The mesh stiffness calculation is based on the Tredgold assumption [21], which says that the straight bevel gears can be approximated as spur gears when projecting on a plane tangent to the back cone. The approximated spur gear will have the pitch radius equal to the back cone, and the pitch will be the same as the bevel gear. A

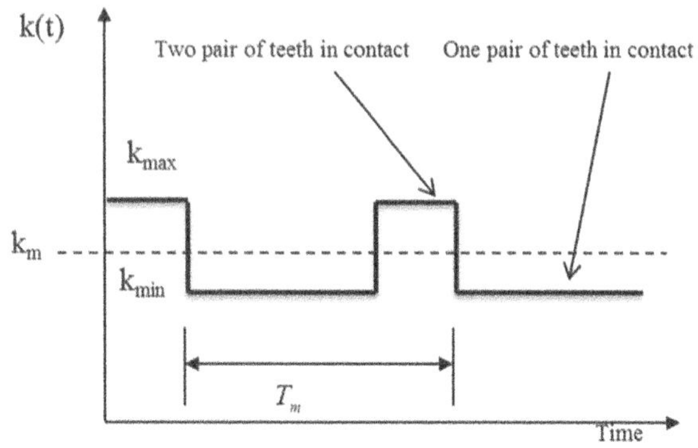

Fig. 2 Model of time-varying mesh stiffness (TVMS)

typical TVMS model of a gear pair is shown in Fig. 2. Here $k(t)$ is the variation of mesh stiffness with respect to time. $k_{\max}$, $k_{\min}$ and k_{m} are maximum, minimum, and mean value of the gear mesh stiffness. T_{m} is the meshing period. The maximum and minimum values of mesh stiffness can be calculated as

$$k_{\min} = k_m\left(1 - \frac{1}{2\varepsilon_\alpha}\right) \tag{12}$$

$$k_{\max} = k_m\left(1 + \frac{2 - \varepsilon_\alpha}{2\varepsilon_\alpha(\varepsilon_\alpha - 1)}\right) \tag{13}$$

where ε_α is the contact ratio and defined as

$$\varepsilon_\alpha = \frac{\sqrt{R_{va1}^2 - R_{vb1}^2} + \sqrt{R_{va2}^2 - R_{vb2}^2} - (R_{v1} + R_{v2})\sin(\alpha)}{\pi m \cos(\alpha)} \tag{14}$$

where R_v R_{va} R_{vb} are the back cone distance, the outer radius of an equivalent spur gear and, base circle radius of an equivalent spur gear respectively. m and α are the module and pressure angle, respectively.

The time evolution of mesh stiffness of a healthy bevel gear pair for two revolutions of the tooth is shown in Fig. 3. The speed of the pinion shaft is taken as $f_p = 7.04\ H_z$. Therefore, the gear mesh frequency obtained is $f_m = 126.7\ H_z$, and the mesh period obtained is $T_m = 0.0079\ s$.

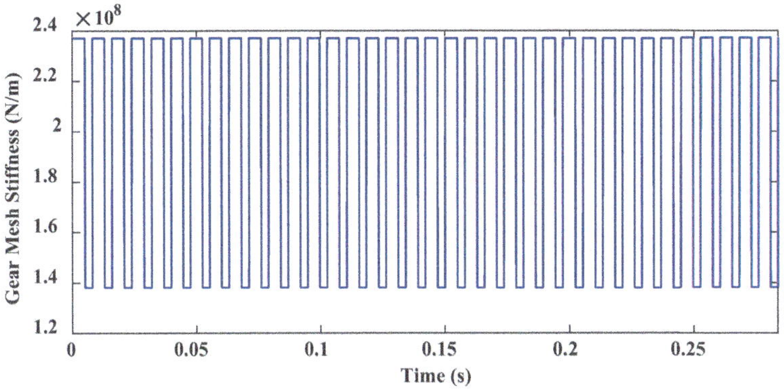

Fig. 3 TVMS of a healthy bevel gear pair

3.2 *Mesh Stiffness Model of Bevel Gear with a Missing Tooth Fault*

A missing tooth fault is considered on the pinion of a straight bevel gear pair. To model the missing tooth fault, an approach proposed by Tian and co-authors [3] is adopted. During the meshing period of teeth, gear mesh stiffness fluctuates between DTPC and STPC for gears (1 < contact ratio < 2). A missing tooth fault can be modeled by assuming there is only STPC in the region of DTPC, and there will be no contact between teeth in the region of STPC for one particular mesh period, i.e., a time period when the missing tooth part (mating part) in a pinion comes in contact with a tooth of a wheel.

The time evolution of mesh stiffness of a straight bevel gear pair with missing tooth fault for two revolutions of a shaft is shown in Fig. 4. The starting point of the meshing is assumed at the position where the missing tooth part of the pinion starts

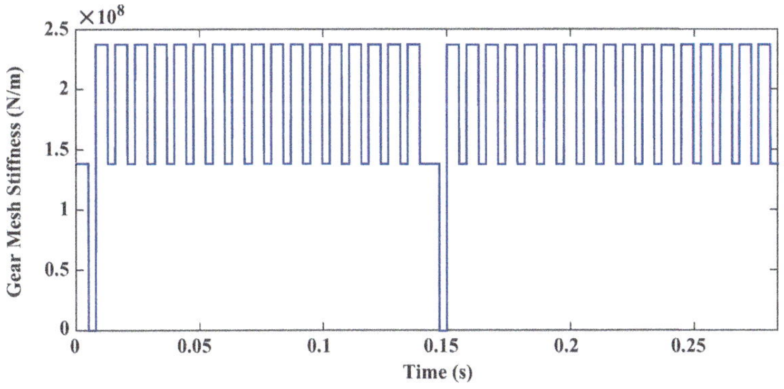

Fig. 4 TVMS of straight bevel gear with a missing tooth fault

mating with the tooth of the wheel. Therefore, for one mesh period of the gears, the contact is assumed to be STPC instead of DTPC, and there is a loss of contact (i.e., zero mesh stiffness) instead of single tooth pair contact. This phenomenon repeats for every revolution of the shaft.

4 Simulation and Results

For simulation purposes, bevel gear parameters of an experimental test rig are considered. The values of the parameters are shown in Table 2. The inertia of the motor and receiver are $I_m = 0.0055$ kg/m^2 and $I_r = 0.1$ kg/m^2.

4.1 Dynamic Response of a Healthy Bevel Gear System

For calculating the dynamic response, as described in Sect. 3.1, the pinion speed is taken as 7.04 Hz, and the corresponding gear mesh frequency is 126.7 Hz. Figure 5a and b shows the time response of a healthy bevel gear computed at the pinion bearing and the wheel bearing respectively, for two rotations of the shaft.

Figure 6a and b shows the time response of a healthy bevel gear computed at the pinion bearing and the wheel bearing respectively for two mesh periods. The time response of healthy bevel gear for two-shaft rotations does not provide any useful information. However, transition regions of STPC and DTPC can be explicitly observed between mesh periods from Fig. 6a and b. Here,T_{DTPC} and T_{STPC} are periods of DTPC and STPC, respectively and the sum of these two time periods is the total mesh period, i.e., $T_{DTPC} + T_{STPC} = T_m$.

Table 2 Parameters of the one-stage straight bevel gear system

Parameters	Pinion	Gear
Module (mm)	2	2
Mean mesh stiffness (N/mm)	2×10^8	
Pressure angle (°)	20°	
Number of teeth	18	27
Mass (kg)	0.03	0.05
Inertia (kg - m^2)	3×10^{-6}	1.15×10^{-5}
Bearing Stiffness (N/m)	$k_x = 1 \times 10^8 k_y = 1 \times 10^8 k_z = 1 \times 10^8$ $k_\phi = 1 \times 10^8$ $k_\psi = 1 \times 10^8$	
Torsional Stiffness (Nm/rad)	$k_{p\theta} = 1 \times 10^4 k_{w\theta} = 2 \times 10^4$	

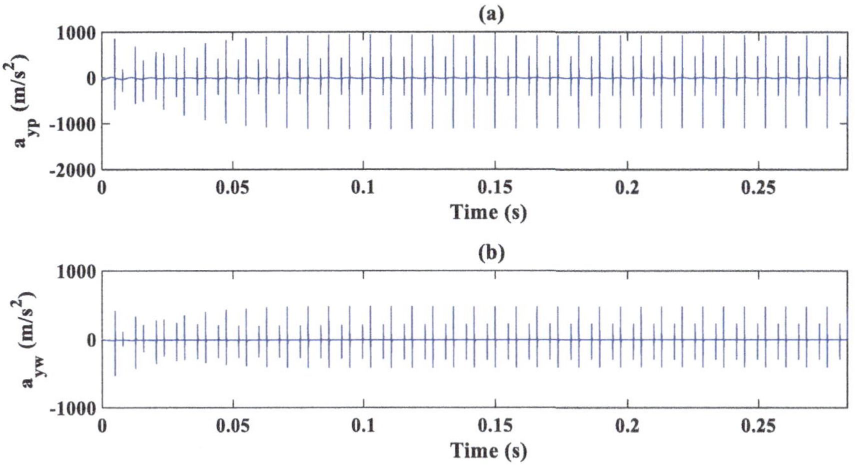

Fig. 5 Time response of healthy bevel gear for two rotations of the shaft at **a** pinion bearing and, **b** wheel bearing

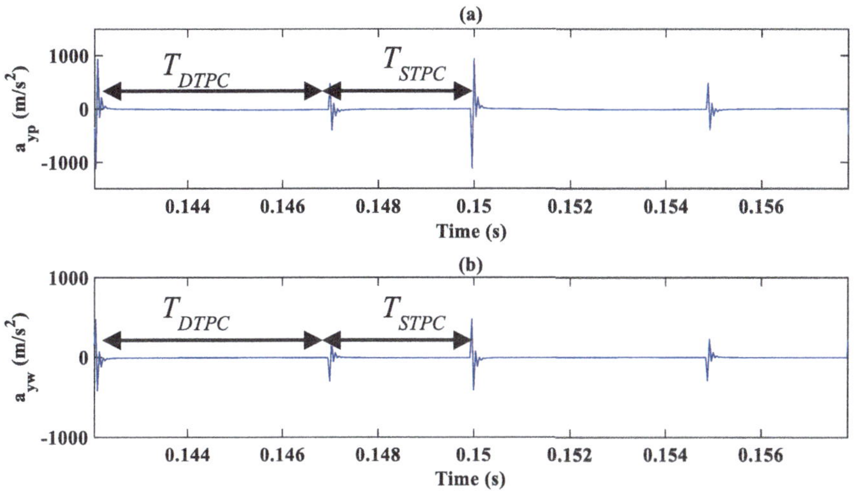

Fig. 6 Time response of healthy bevel gear for two mesh periods at **a** pinion bearing and, **b** wheel bearing

4.2 Dynamic Response of a Bevel Gear System with Missing Tooth Fault

The time response of one stage straight bevel gear system with missing tooth fault computed at the pinion bearing and the wheel bearing respectively for two rotations of the shaft is shown in Fig. 7a and b respectively. For two mesh periods, the time

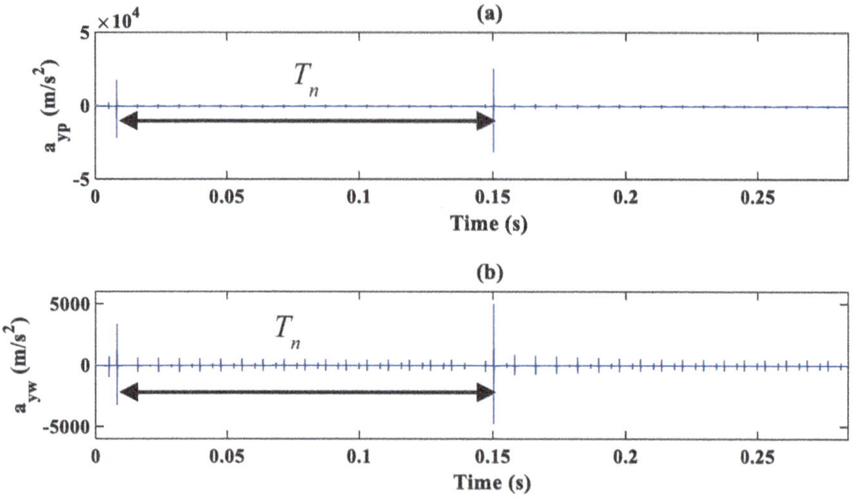

Fig. 7 Time response of the bevel gear with missing tooth fault for two rotations of the shaft at **a** pinion bearing and, **b** wheel bearing

response computed at the pinion bearing and the wheel bearing is shown in Fig. 8a and b, respectively. In Fig. 7a and b, the impact due to missing tooth fault can be seen at every rotation of the shaft when the wheel mates with missing tooth part on the pinion. The corresponding accelerations are also higher compared to that of a healthy one. Here T_n is the period of one rotation of pinion shaft.

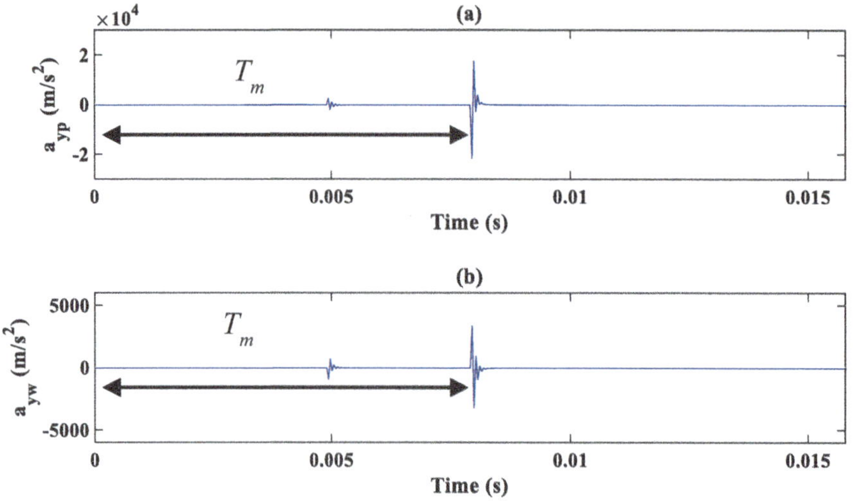

Fig. 8 Time response of the bevel gear with missing tooth fault for mesh periods at **a** pinion bearing and, **b** wheel bearing

Similarly, a higher value of impact is also observed in Fig. 8a and b after each mesh period compared to that of a healthy one. It can be noted that the higher acceleration values (order of 10^4) obtained only at pinion bearing, in contrast to the wheel bearing, because the fault has been introduced in the pinion.

5 Experimental Validation

For the validation of simulated results, an experiment was conducted on a single-stage straight bevel gearbox with a healthy and missing tooth fault case. A bevel gearbox mounted on machinery fault simulator was used as a test rig (see Fig. 9). The healthy bevel gear and bevel gear with a missing tooth fault used in the experiment are shown in Fig. 10a and b, respectively.

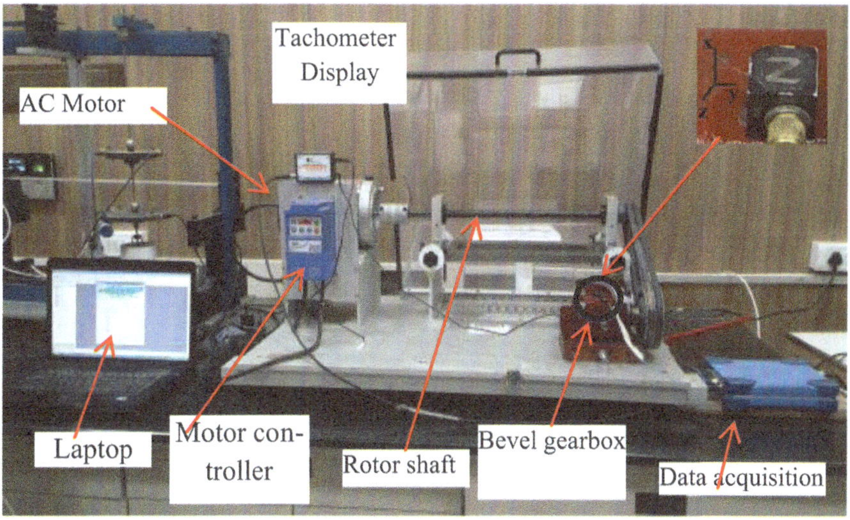

Fig. 9 Experimental test-rig with a zoomed view of the accelerometer

Fig. 10 Bevel bears with different health conditions **a** healthy and, **b** with missing tooth

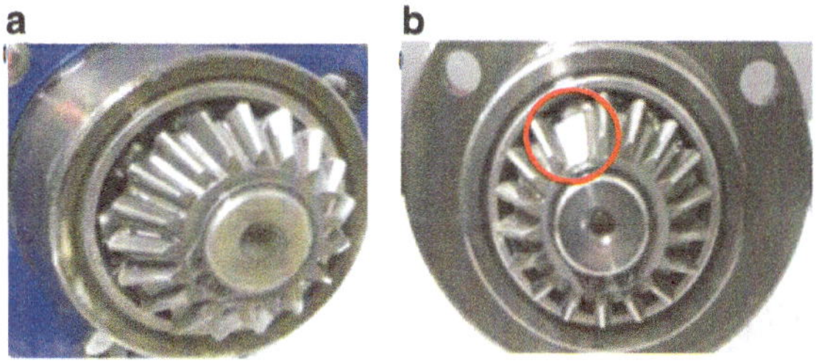

The vibration signals are acquired using the tri-axial accelerometer. The motor speed is chosen as 18 Hz and after reduction through a belt-drive, the gearbox input (pinion) shaft speed obtained was 7.04 Hz. The input speed was measured using a tachometer. The acceleration signals were acquired at a sampling rate of 6.4 kHz. The acceleration measurements were done in all three directions at pinion bearing, as shown in the zoomed view of accelerometer in Fig. 9. The specification of the gearbox is shown in Table 3. After acquiring time-domain signals, it was then processed in MATLAB to obtain the Fourier transform of the signal.

The experimental time response of a healthy gearbox for two rotations of the shaft shown in Fig. 11a does not provide any useful information. However, if the response is observed for two mesh periods shown in Fig. 11b, the transition of DTPC and STPC can be inferred as an increase in amplitude. The simulated frequency response of the healthy bevel gear at pinion bearing is shown in Fig. 12. The gear mesh frequencies and its harmonics in the frequency domain can be explicitly observed. The experimental frequency response of a healthy bevel gear is shown in Fig. 13. The gear mesh frequency and some of its harmonics are observed.

The responses for bevel gear with missing tooth fault are acquired using the same input conditions and sample rates that were used in the case of healthy bevel gears. The time response with missing tooth fault for two rotations of the shaft and two mesh periods are shown in Fig. 14a and b, respectively. From Fig. 14a, an impact caused by missing tooth fault at each rotation of the shaft is observed. The acceleration value in Fig. 14a is also higher than that of a healthy one. Figure 14b does not provide any

Table 3 Specifications of the bevel gear

Parameters	Value
Gear ratio	1.5:1
Wheel pitch angle	56°19′
Pinion pitch angle	33°41′
Pressure angle for wheel and pinion	20°
Number of teeth in pinion	18
Number of teeth in the wheel	27

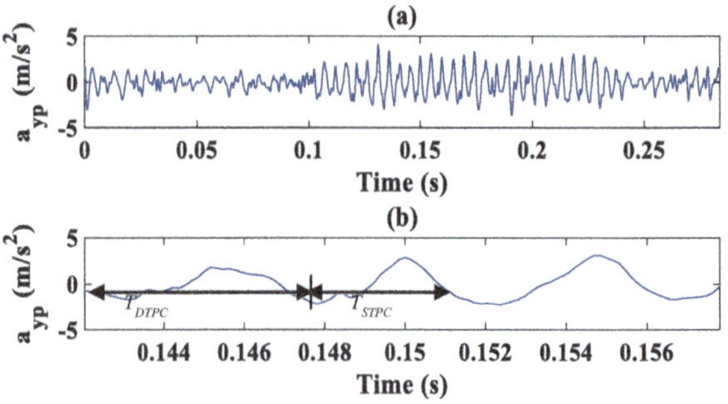

Fig. 11 Experimental time response of healthy bevel gear at pinion bearing **a** for two rotations of the shaft. **b** For two mesh periods

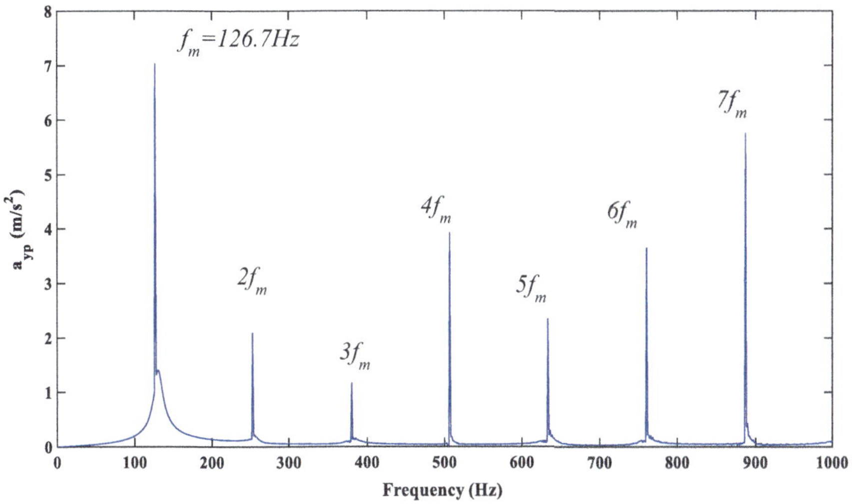

Fig. 12 The frequency response of healthy bevel gear at pinion bearing

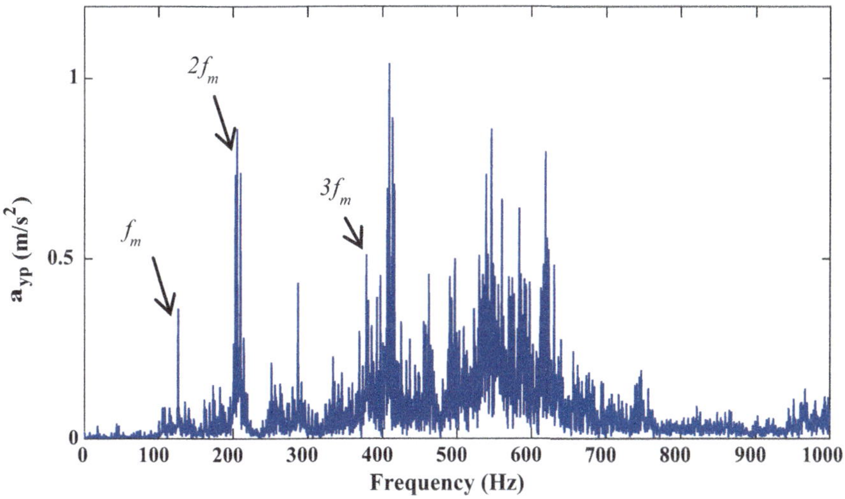

Fig. 13 The experimental frequency response of healthy bevel gear at pinion bearing

clear information about the impact. However, the measurement of two crests equal to the mesh period suggests the variation in the amplitude at the transition region.

Figure 15 shows the simulated frequency response at pinion bearing with missing tooth fault. In the frequency domain, the fault frequencies appear as the sidebands around the meshing frequency as $f_m \pm f_p$ where f_p is the frequency of the pinion. The sidebands due to fault also appear around the harmonics of the gear mesh frequency as $2f_m \pm 2f_p$, $3f_m \pm 3f_p$ and so on.

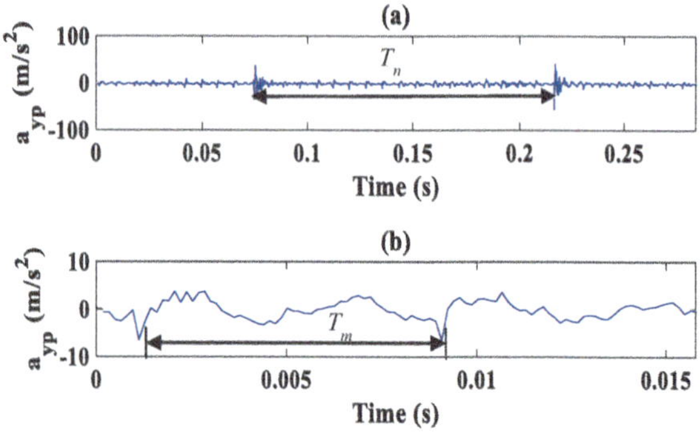

Fig. 14 Experimental time response of bevel gear with the missing tooth at pinion bearing **a** for two rotations of the shaft. **b** For two mesh periods

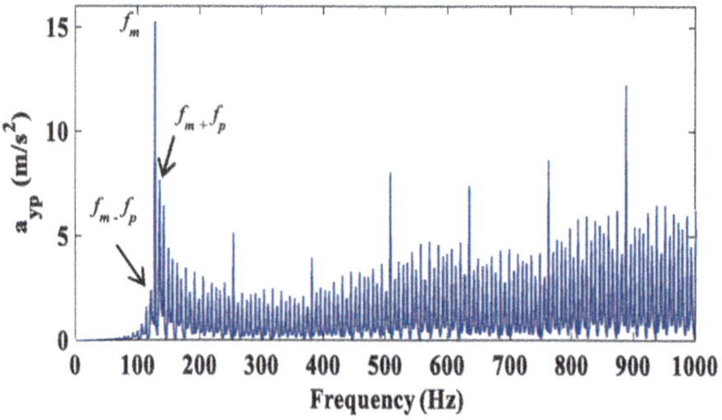

Fig.15 Simulated frequency response of bevel gear with missing tooth fault at pinion bearing

The experimental frequency response of the bevel gearbox with missing tooth fault at pinion bearing is shown in Fig. 16. The fault frequencies around sidebands equal to $2f_m \pm 2f_p$ are observed. Other harmonics of the fault frequencies can also be observed from Fig. 16 but not indicated. However, due to some reason $f_m \pm f_p$ is not appearing in the experimental response.

Figure 17 shows the zoomed view of the simulated frequency response (see Fig. 15), where gear mesh frequency and its second harmonic can be seen clearly

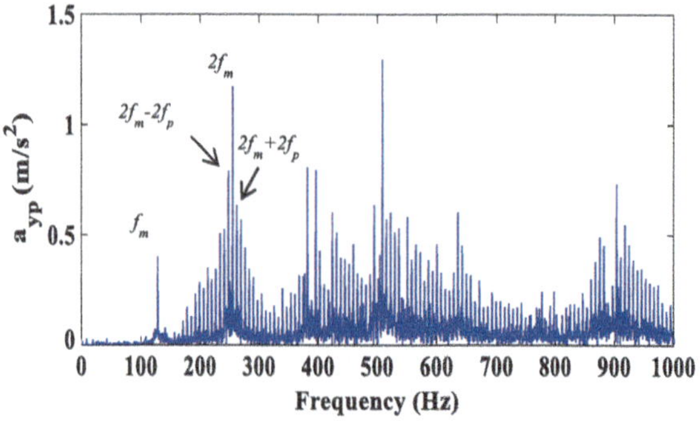

Fig. 16 The experimental frequency response of bevel gear with the missing tooth at pinion bearing

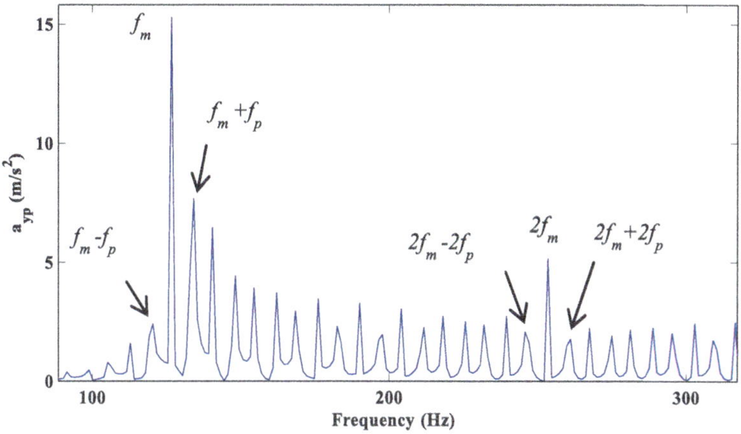

Fig. 17 Zoomed view of the simulated frequency response of bevel gear with missing tooth fault at pinion bearing

with sidebands around them. A zoomed view of the experimental frequency response, shown in Fig. 16, is presented in Fig. 18.

Comparison of second harmonic of Fig. 17 with Fig. 18 yields a good agreement between simulation and experimental results. Also, a comparison between time responses of Figs. 8a and 14a show good agreement between simulation and experimental results for missing tooth fault.

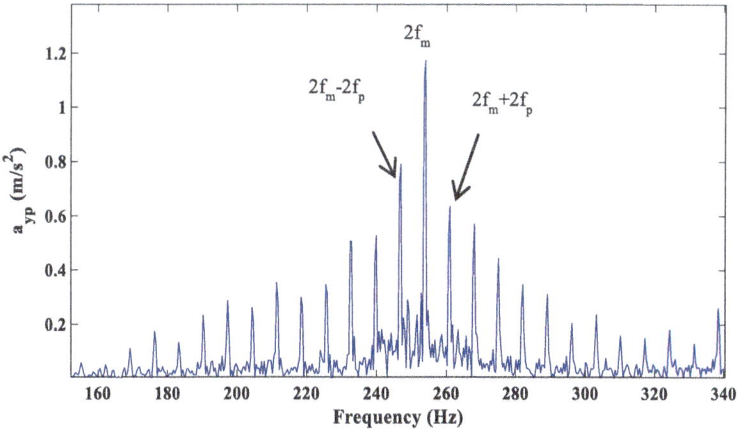

Fig. 18 Zoomed view of the experimental frequency response of bevel gear with missing tooth fault at pinion bearing

6 Conclusion

In the present study, a gear mesh stiffness model of bevel gear with missing tooth fault is proposed using square wave approximation. The mesh stiffness model is incorporated into the dynamic model of a one-stage straight bevel gearbox, and the dynamic response is calculated. The dynamic characteristics due to missing tooth fault are described as follows:

- The impacts caused by missing tooth faults are observed in time response of the simulated signal at each shaft rotation. Some distinct characteristics can also be inferred from the time response between mesh periods, for example, the transition of teeth pairs.
- The fault frequencies are observed as sidebands around the gear mesh frequencies in the simulated frequency response.

To validate the simulation results, experiments are performed using a test-rig having a straight bevel gearbox mounted on it. The response is taken for both the healthy and the missing tooth fault cases. The experimental results agree well with the simulation results both in time and frequency domain. Therefore, it is concluded that the proposed model successfully identifies the fault in a single stage straight bevel gearbox.

References

1. Randall RB (1982) A new method of modeling gear faults. Trans Asme J Mech Des 104:259–267. https://doi.org/10.1115/1.3256334
2. Chaari F, Baccar W, Abbes MS, Haddar M (2008) Effect of spalling or tooth breakage on gearmesh stiffness and dynamic response of a one-stage spur gear transmission. Eu J Mech A/Solids 27:691–705. https://doi.org/10.1016/j.euromechsol.2007.11.005
3. Tian X, Zuo MJ, Fyfe KR (2004) Analysis of the vibration response of a gearbox with gear tooth faults. In: Proceedings of IMECE04 2004 ASME international mechanical engineering congress and exposition. 13–20 Nov 2004, Anaheim, Calif. USA. 785–793. doi:https://doi.org/10.1115/imece2004-59224
4. Bruyère J, Dantan JY, Bigot R, Martin P (2007) Statistical tolerance analysis of bevel gear by tooth contact analysis and Monte Carlo simulation. Mech Mach Theory 42:1326–1351. https://doi.org/10.1016/j.mechmachtheory.2006.11.003
5. Peng T, Lim TC (2018) Influence of gyroscopic effect on hypoid and bevel geared system dynamics. SAE Int J Passeng Cars-Mech Syst 2:1377–1386
6. Feng Z, Wang S, Lim TC, Peng T (2011) Enhanced friction model for high-speed right-angle gear dynamics. J Mech Sci Technol 25:2741–2753. https://doi.org/10.1007/s12206-011-0803-3
7. Chang-Jian CW (2011) Nonlinear dynamic analysis for bevel-gear system under nonlinear suspension-bifurcation and chaos. Appl Math Model 35:3225–3237. https://doi.org/10.1016/j.apm.2011.01.027
8. Peng T, Lim TC, Yang J (2011) Eccentricity effect analysis in right-angle gear dynamics. In: Proceedings of ASME 2011 international design engineering technical conference and computer and information in engineering conference. 1–14

9. Yinong L, Guiyan L, Ling Z (2010) Influence of asymmetric mesh stiffness on dynamics of spiral bevel gear transmission system. Math Probl Eng 1–13. doi:https://doi.org/10.1155/2010/124148
10. Liang X, Zuo MJ, Feng Z (2018) Dynamic modeling of gearbox faults: a review. Mech Syst Signal Process 98:852–876. https://doi.org/10.1016/j.ymssp.2017.05.024
11. Karray M, Chaari F, Viadero F, del Rincon AF, Haddar M (2013) Dynamic response of single stage bevel gear transmission in presence of local damage. New trends mech mach sci vol 7, pp 337–345. doi:https://doi.org/10.1007/978-94-007-4902-3
12. Driss Y, Hammami A, Walha L, Haddar M (2014) Effects of gear mesh fluctuation and defaults on the dynamic behavior of two-stage straight bevel system. Mech Mach Theory 82:71–86. https://doi.org/10.1016/j.mechmachtheory.2014.07.013
13. Karray M, Chaari F, Khabou MT, Haddar M (2018) Dynamic analysis of bevel gear in presence of local damage in nonstationary operating conditions. In: Proceedings of 7th conference design and modeling of mechaical systems. C. 27–29 Mar 2018, Hammamet, Tunis, 325–330. doi:https://doi.org/10.1007/978-3-319-66697-6-32.
14. Ramteke DS, Parey A, Pachori RB (2019) Automated gear fault detection of micron level wear in bevel gears using variational mode decomposition. J Mech Sci Tech 33(12):5769–5777. https://doi.org/10.1007/s12206-019-1123-2
15. Lafi W, Djemal F, Tounsi D, Akrout A, Walha L, Haddar M (2019) Dynamic modelling of differential bevel gear system in the presence of a defect. Mech Mach Theory 139:81–108. https://doi.org/10.1016/j.mechmachtheory.2019.04.007
16. Park M (2003) Failure analysis of an accessory bevel gear installed on a J69 turbojet engine. Eng Fail Anal 10:371–382. https://doi.org/10.1016/S1350-6307(02)00071-7
17. Bel Mabrouk I, El Hami A, Walha L, Zghal B, Haddar M (2017) Dynamic response analysis of Vertical Axis Wind Turbine geared transmission system with uncertainty. Eng Struct 139:170–179. https://doi.org/10.1016/j.engstruct.2017.02.028
18. Gui Y, Han QK, Li Z, Chu FL (2014) Detection and localization of tooth breakage fault on wind turbine planetary gear system considering gear manufacturing errors. Shock and Vib 692347:13. https://doi.org/10.1155/2014/692347
19. Brethee KF, Gu F, Ball AD (2016) Frictional effects on dynamic response of gear systems and the diagnostics of tooth breakage. Sys Sci Cont Eng 4:270–284. https://doi.org/10.1080/21642583.2016.1241728
20. Yang W, Jiang D, Han T (2017) Effects of tooth breakage size and rotational speed on the vibration response of planetary gearbox. App Sci 7(7):678. https://doi.org/10.3390/app7070678
21. Elkholy AH, Elsharkawy AA, Yigit AS (1998) Effect of meshing tooth stiffness and manufacturing error on the analysis of straight bevel gears. J Struct Mech 26(1):41–61. https://doi.org/10.1080/08905459808945419

Investigating the Electro-mechanical Interaction Between Helicoidal Gears and an Asynchronous Geared Motor

Safa Boudhraa, Alfonso Fernandez del Rincon, Mohamed Amine Ben Souf, Fakher Chaari, Mohamed Haddar, and Fernando Viadero

Abstract Gears are widely used in different domains, aeronautics, automotive and machines tools manufacturing etc. ... Therefore, monitoring these mechanical systems have been a huge scientific and industrial trend recently. Different studies were orientated to its investigating using different techniques in order to study the sensitivity of each. Motor Stator Current Analysis (MSCA) was one of the highlighted techniques for its easy accessibility and accuracy. The mechanical system generates torque oscillations which leads to a frequency and amplitude modulation effects on the stator current in the asynchronous machine. Within this context, this study was done on an experimental test bench composed of a geared motor and a helicoidal gears. The current signal was recorded using clamp meter and presented on LMSTestLab. The electromagnetic interaction starts to get visible in the mechanical frequencies seen in the sidebands around the motor frequency. In this paper, some of the results would be present to illustrate the electro-mechanical interaction between the mechanical system and the electrical part, which is itself composed of an asynchronous motor and an integrated gearbox.

Keywords Geared motor · Helicoidal gears · Stator current · Electromechanical interaction

1 Introduction

The gearboxes are widely used in different industrial domains such as automotive, aeronautics and machinery manufacturing tools. Hence, because of its critical role, the condition monitoring of these mechanisms has been in permanent progress by

S. Boudhraa (✉) · M. A. B. Souf · F. Chaari · M. Haddar
Laboratory of Mechanics, Modelling and Production (LA2MP), National School of Engineers of Sfax, BP1173, 3038 Sfax, Tunisia

S. Boudhraa · A. F. del Rincon · F. Viadero
Department of Structural and Mechanical Engineering, Faculty of Industrial and Telecommunications Engineering, Avda de Los Castros S/N 39005, University of Cantabria, Santander, Spain

F. Chaari et al. (eds.), *Smart Monitoring of Rotating Machinery for Industry 4.0*, Applied Condition Monitoring 19, https://doi.org/10.1007/978-3-030-79519-1_8

involving different physical phenomena. Researchers had been orientated to experimental tests also to developing numerical models in order to facilitate the detection of any anomaly, localize it and mainly to avoid the high price of mounting, dismounting and manufacturing. Over the past two decades, using the vibration signals dominate the condition monitoring of rotating machinery in different operating conditions [1, 2].

However, the stator motor current is one of the leading methods in this field today, mainly for its easy accessibility. Such as Kia et al. [3] who had referred to the impact of the torsional vibrations on the stator current. In a more advanced work Ottewill et al. [4] in their paper had worked on monitoring a tooth defect in epicyclic gearboxes using numerical modelling pursued by experimental validation.

It has been shown also [5, 6], that due to the torsional vibration resulted from the load oscillation in the output wheels and the stiffness variation of the gear teeth contact, the gearbox adds the rotation and mesh frequency components into the torque signature. By means, this impact makes the stator current multi-component phase modulated. The amplitude of each gearbox-related frequency component in the stator current spectrum depends on its respective modulation index value. In [7, 8] Kia et al. found that the rotating frequencies related to the motor can be clearly seen in the current spectrum for different amplitudes. This impact was first studied on a single frequency effect on the current signal by Yacamini [9] and developed later by Kar et al. [10] on a multistage of gears. Within this context, this work aims to present the electro-interaction between a geared motor and a stage of helicoidal gears.

This paper is structured as followed, besides to the introduction, a second section to describe the experimental system composed of a pair of helicoidal gears driven by a geared motor. Also, describing the emplacement of the sensor along the test bench. In the third section, different experimental measurements are presented in both time domain and frequency domain. Finally, a fourth section states conclusions beyond the previous work.

2 Experimental Set Up

The objective of this paper is to investigate the impact of different mechanical components connected to an electrical machine on its behaviour. Therefore, this study is accomplished on an experimental test bench as illustrated in Fig. 1.

The test bench is composed of two parts:

- A geared motor: a Bauer geared motor (Type: BG50-11D099A4-Tk-K311) shown in the Fig. 2. The driving system is composed by a three-phased asynchronous motor (4 poles, 1.1 kW, Fe = 50 Hz, 1400 rpm) and a double-stages gearbox. Both stages are composed of helicoidal gears (Table 1)
- A pair of helicoidal gears: ($Z_1 = 60$, and $Z_2 = 30$) where Z_i is the number of gear's teeth.

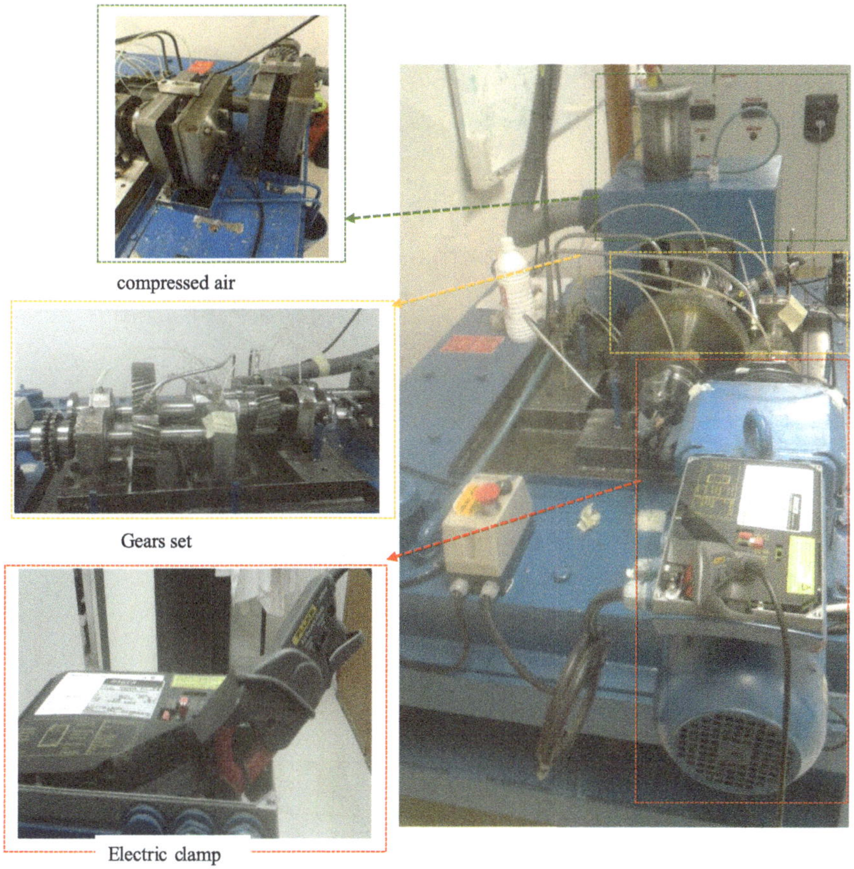

Fig. 1 The experimental test rig

Besides the motor and gears, the load is introduced by a compressed air brake connected the output shafts as explained in the Fig. 3. Also, in order to record the electrical signal of we used an electrical clamp connected to LMSTestLab. Later all the results will be plotted and analysed on MATLAB.

For a deep qualitative study, the system in monitored to 840 rpm where all the key frequencies of the system are presented as followed in Table 2

3 Results

Two configurations were taken into consideration in this study. A first one illustrated in Fig. 4, is studying the current signals for a free motor in order to establish a

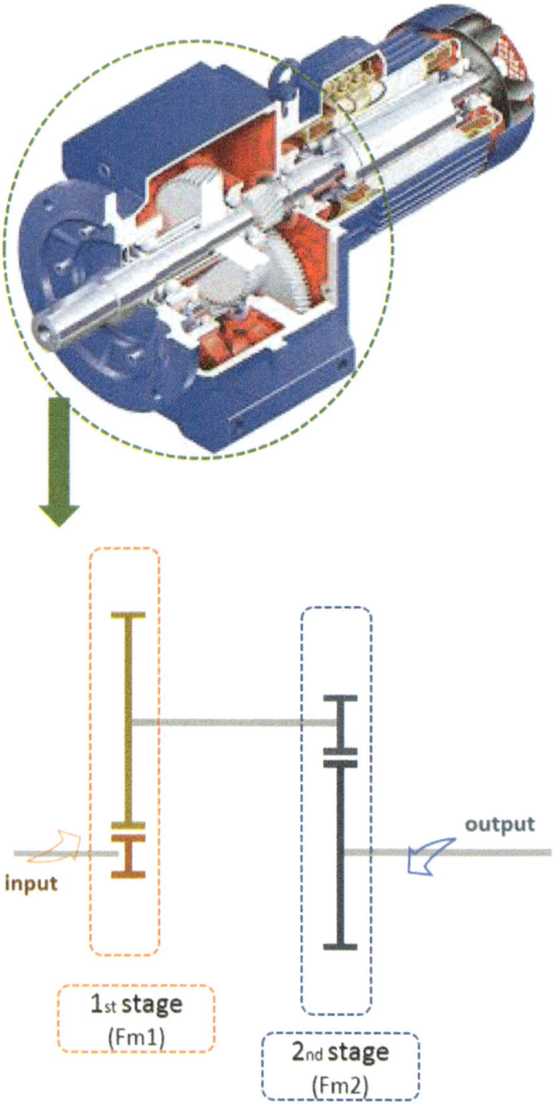

Fig. 2 The geared motor structure

Table 1 The gears characteristics

	Teeth number		Reduction ratio
	First stage	Second stage	
The motor	$Z_{11} = 10$	$Z_{21} = 62$	47.02
	$Z_{12} = 91$	$Z_{22} = 12$	
Mechanical system	$Z_1 = 60$		2
	$Z_2 = 30$		

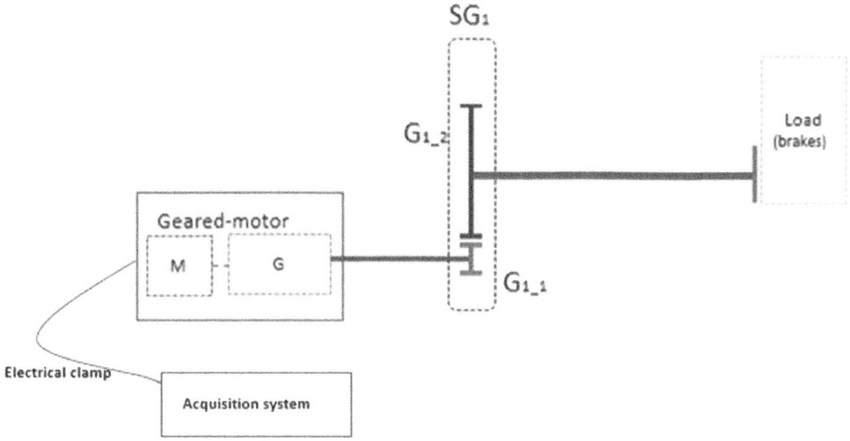

Fig. 3 The global schema of the test bench

Table 2 Key Frequencies

		Key frequency			
The motor	F_e= 30 Hz	1st stage	14	1.3	141.12
		2nd stage	2.6	0.3	18.6
Mechanical system			0.3	0.1	6

F_{gi} is the frequency of the ith gear, F_{mi} is the meshing frequency of the ith stage in the motor, F_e is the electrical frequency

reference situation and identify all the electrical system related frequencies in the current frequency spectrum.

- **Free motor**:

After applying the Fast Fourier Transform on the temporal signal, the Fig. 5a illustrates the stator current spectrum for a free motor. It is seen that the current spectrum is dominated by the supply frequency given by 50 Hz. With lateral sidebands related the mechanical rotating frequency. Meanwhile in order to accentuate the appearance of the mechanical system's signature in the measured current Fig. 5b showing the same results in the logarithm scale. It is totally seen the appearance of additional frequency besides to the previous one explained.

Figure 6 presents the current frequency spectrum with emphasizing the electromechanical contact.

It is seen in Fig. 6 that the connection of the gearbox to the motor impacts the current signal by the presence of the gear meshing frequency of each stage. Whenever there is a load fluctuation, a change in speed occurs thus changing the per unit slip, which subsequently causes changes in sidebands across the line frequency (fe). Figure 6c presents highlights the appearance of the mechanical impact of the gears

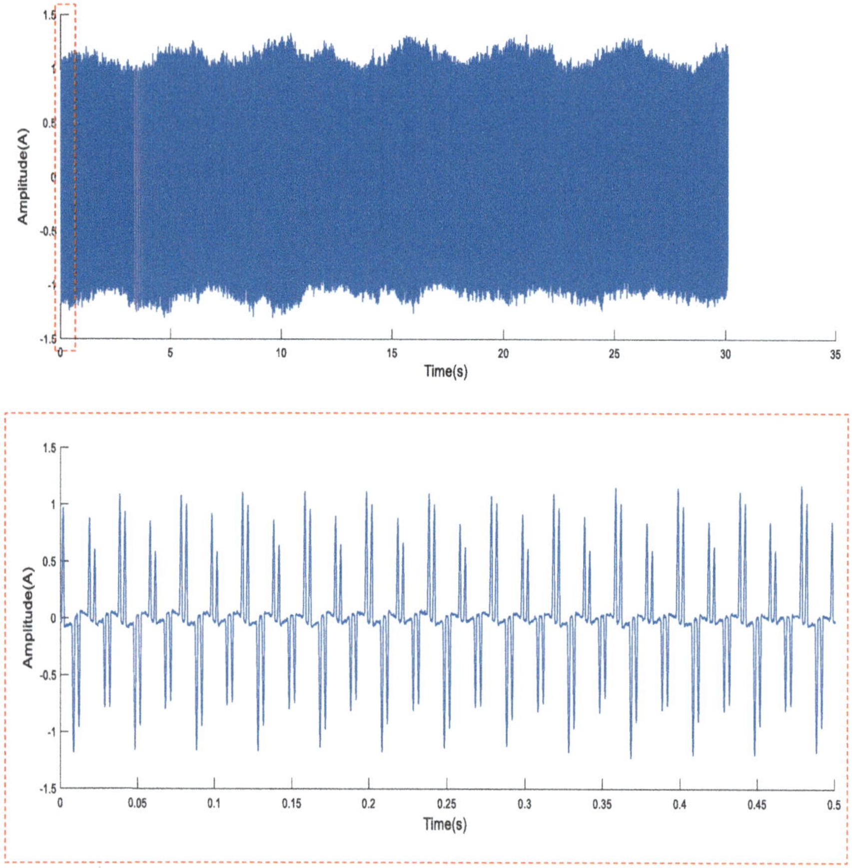

Fig. 4 The current signal

through the gear meshing frequency. The peaks seen in the figure are presenting the relate gear meshing frequency as given in the expression:

$$f_{ri} = |f_e \pm m f_{mi}| \tag{1}$$

where: i refers to whether the 1st or the 2nd stage of the integrated gearbox and $m \in N$.

Even though, with a weak amplitude, even related mechanical frequencies can be seen in the Fig. 6b, and those frequencies are given by:

$$\left| f_e \pm m f_{m1} \pm n f_{m2} \pm p f_{g1} \right| \tag{2}$$

where:$m, n, p \in N$.

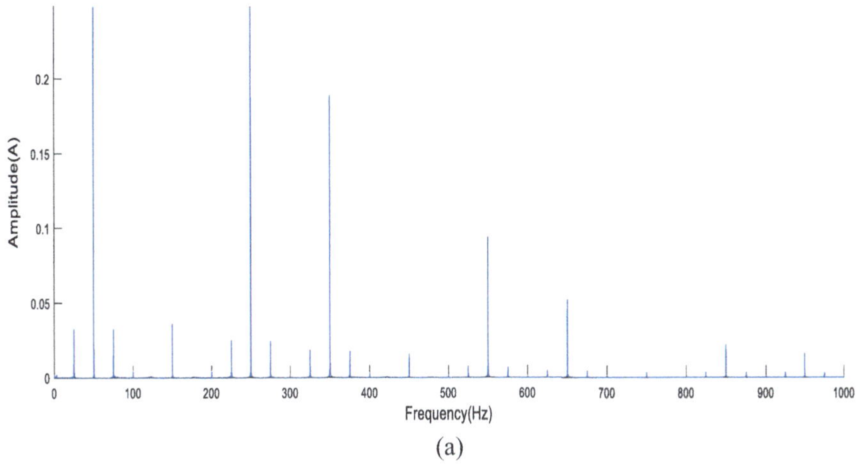

Fig. 5 The frequency spectrum of the current signal for free motor **a** linear scale, **b** logarithmic scale

- **Connected to the gears**:

Figure 7 presents a brief compaison between the frequency spectrum of the current signal for free motro (in blue) and for a motor connected to the mechanical system, pair af helicoidal gears (in green). It is tottaly freseen the appearance of additional frequencies related to the rotating frequencies and the meshing frequency as well.

F_{ms1} is the meshing frequency of the mechanical system. Figure 7b presents the appearance of peaks in the significant frequencies $F_e + F_{ms1}$ and $F_e + 2F_{ms1}$. In fact, the illustration of this last measurement is for objective to emphasize the sensitivity of the electrical signals in mechanical connections detection as shown in Fig. 7c.

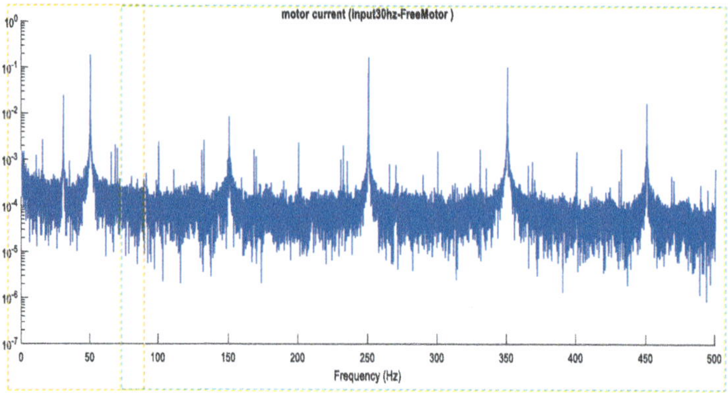

(a)

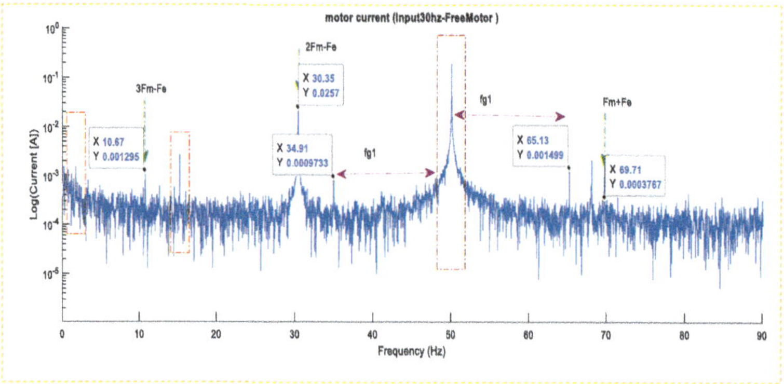

(b)

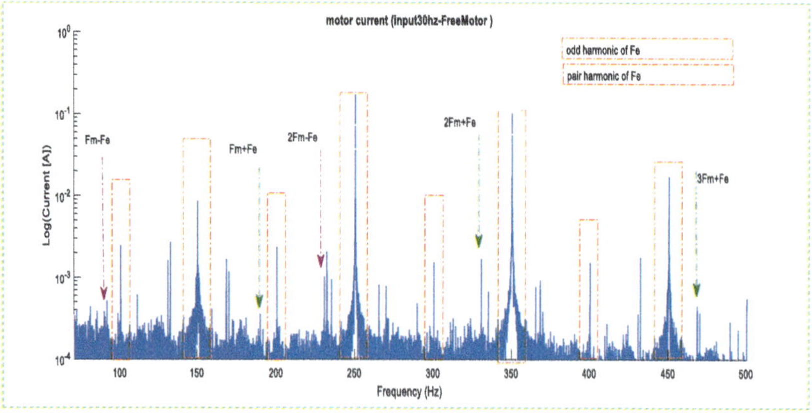

(c)

Fig. 6 The current frequency spectrum highlighting the impact of the gearbox connected to the motor

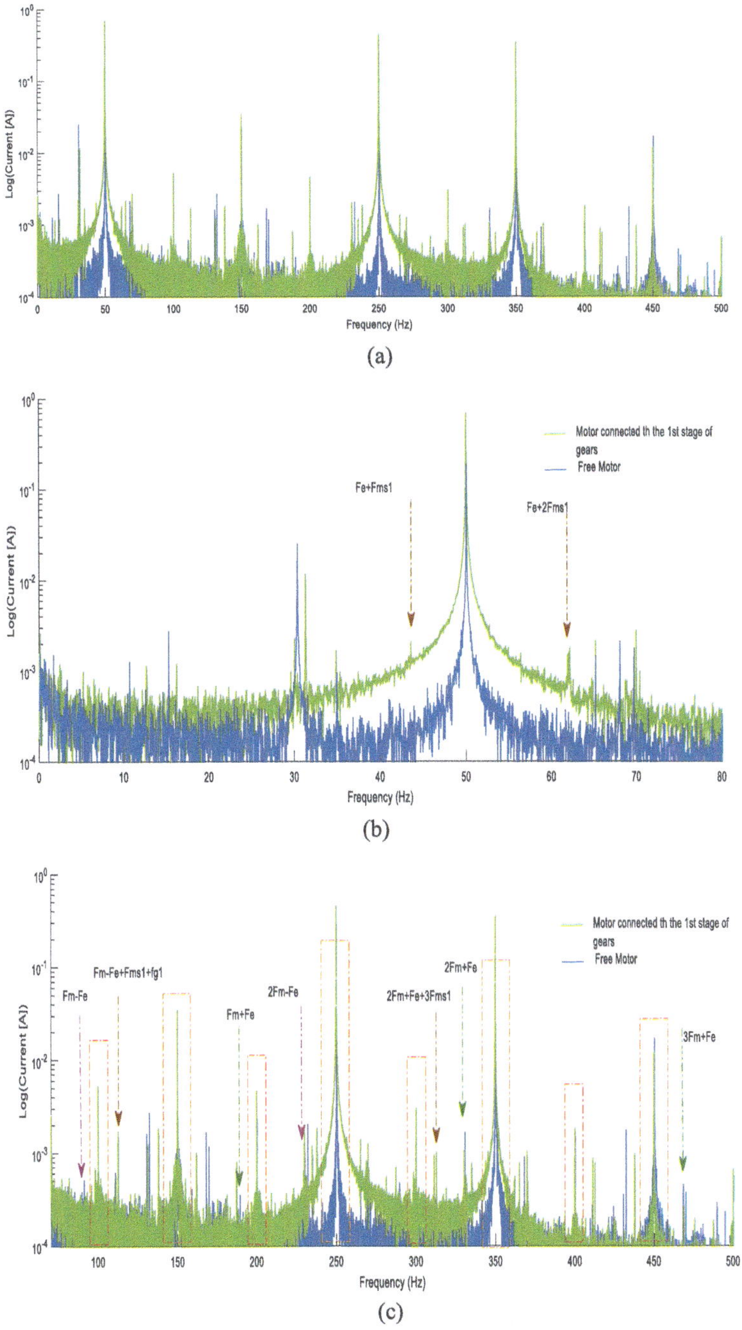

Fig. 7 The current spectrum while the motor is connected to the helicoidal pair of gears

Also, it is seen the increase of the amplitude of all the peaks related the related-mechanical frequencies which was related to the high load introduced for the second configuration.

4 Conclusion

The motor current gives a significant proof of the electromechanical interaction between the mechanical system, seen in this work as a gearbox and the asynchronous motor. In this paper we studied the impact of the gears' contact on the electrical system by analysing the current signals obtained experimentally. These measurements were at first place treated using FFT to transform the averaged time signal in to the frequency domain, in order to highlight the presence of the mechanical impact.

The motor used is already connected to an integrated gearbox therefore we explain the appearance of related mechanical frequencies in the current spectrum even for a free motor configuration. These results were more visible and highlighted after connected the motor to a couple helicoidal gears. For coupled motor to the gearbox, we noticed the appearance of additional related frequency in the current spectrum. In fact, the appearance of the gear meshing frequency is translating the impact of the load fluctuation due to the contact between teeth on the current signal. Hence, for a loaded system it is more seen the impact of the mechanical contacts whether due to the meshing phenomena or simply to gears rotation. This paper is a reface for future investigation which will study the impact of different conditions on the monitoring of gearboxes.

Acknowledgements The authors would like also to acknowledge the help provided by the project "Dynamic behaviour of gear transmissions in non-stationary conditions", ref. DPI2017-85390-P, funded by the Spanish Ministry of Science and Technology. We would like to thank the University of Cantabria cooperation project for the doctoral training to Sfax University students.

References

1. Razafimahefa TD, Sambatra EJR, Heraud N (2016) Study of the evolution of an inter turns short circuit fault in induction machine. In: 2016 international conference and exposition on electrical and power engineering (EPE), pp 179–184. IEEE
2. Wakileh GJ (2003) Harmonics in rotating machines. Electric Power Syst Res 66(1):31–37
3. Kia SH, Henao H, Capolino GA (2007) Gearbox monitoring using induction machine stator current analysis. In: IEEE international symposium on diagnostics for electric machines, power electronics and drives (SDEMPED 2007). IEEE, pp 149–154
4. Ottewill JR, Ruszczyk A, Broda D (2017) Monitoring tooth profile faults in epicyclic gearboxes using synchronously averaged motor currents: mathematical modeling and experimental validation. Mech Syst Signal Process 84:78–99
5. Chen K, Hu J, Peng Z (2017) Analytical framework of gearbox monitoring based on the electro-mechanical coupling mechanism. Energy Procedia 105:3138–3145

6. Blodt M, Chabert M, Regnier J, Faucher J (2006) Mechanical load fault detection in induction motors by stator current time-frequency analysis. IEEE Trans Ind Appl 42(6):1454–1463
7. Benbouzid MEH, Vieira M, Theys C (1999) Induction motors' faults detection and localization using stator current advanced signal processing techniques. IEEE Trans Power Electron 14(1):14–22
8. Kliman GB, Stein J (1992) Methods of motor current signature analysis. Electric Mach Power Syst 20(5):463–474
9. Yacamini R, Smith KS, Ran L (1998) Monitoring torsional vibrations of electro-mechanical systems using stator currents. J Vibr Acoustics 120(1):72–79
10. Kar C, Mohanty AR (2006) Monitoring gear vibrations through motor current signature analysis and wavelet transform. Mech Syst Signal Process 20(1):158–187

Algebraic Estimator of Damping Failure for Automotive Shock Absorber

Maroua Haddar, Riadh Chaari, S. Caglar Baslamisli, Fakher Chaari, and Mohamed Haddar

Abstract One of the challenges for automotive industry is online fault identification and elimination of vehicle interior vibration. The effectiveness of semi-active shock absorbers can be threatened by additive and multiplicative perturbations. In fact, these perturbations are able to cripple the dynamics and complicate shock absorber operation. In particular, the ride comfort criteria should be insensitive to unpredictable troubles. Moreover, mechanical systems must operate in healthy conditions. Getting online-information about failures can be achieved by a simple algebraic estimator. These algebraic tools are based on operational calculus. The algebraic estimator has the fastest detection time and non-asymptotic behavior. In literature, the algebraic observers shown better robustness to vehicle mass uncertainties. Additionally, this estimator requires a lower number of sensors and has a lower computational overhead. Motivated by the above analysis, a simple quarter car model was used to test this approach. A restricted model is sufficient for identifying the hysteresis behavior of Magneto rheological damper. Only a sliding window needs to be tuned by the operator for obtaining a good estimation. Furthermore, just vehicles displacements are needed for identifying damping force online. The numerical simulations illustrate the effectiveness of proposed identification tool under different kind of perturbations.

Keywords Online identification · Semi-active · Hysteresis · Monitoring

M. Haddar (✉) · R. Chaari · F. Chaari · M. Haddar
Mechanics, Modeling and Production Laboratory (LA2MP), Mechanic Department, National Engineering School of Sfax (ENIS), 1173-3038 Sfax, BP, Tunisia
e-mail: maroua.haddar@enis.tn

M. Haddar
e-mail: Mohamed.haddar@enis.rnu.tn

S. C. Baslamisli
Department of Mechanical Engineering, Hacettepe University, 06800 Beytepe, Ankara, Turkey

F. Chaari et al. (eds.), *Smart Monitoring of Rotating Machinery for Industry 4.0*,
Applied Condition Monitoring 19, https://doi.org/10.1007/978-3-030-79519-1_9

1 Introduction

Semi-active suspensions get the interest of automotive designers because it uses low energy at low cost compared to active controllers. The control power depends greatly on the choice of the appropriate control schemes of the adaptable shock absorber. A Magneto-Rheological damper (MR) that contains smart fluid with magnetic particles surrounded in a type of synthetic oil is used.

Wider range of semi-active controllers implements condition monitoring devices to diagnose unpredictable failures that affect the effectiveness of MR damper. There are different forms of nonlinearities that can affect the dynamic behaviour of vertical motion of suspension system. However, in real application, these nonlinearities are present but the knowledge of precise mathematical model of this kind of perturbation is difficult.

Researchers have proposed several solutions to enhance the semi-active suspension system performance. Fault Detection and Isolation [1], the Unknown Input Observer [2] and a pseudo inverse actuator estimation [3] are different techniques for identifying semi-active actuator defects. Most of these strategies are based on vehicle model and asymptotic observer. Based on the previous analysis, this paper suggests an online algebraic estimator to estimate the damping force of MR damper including its abnormale behavior. Based on algebraic and operational rules, a restricted model for monitoring the health of damping force online is formulated.

The choice of the differential–algebraic theory for estimation is based on characteristic features of finite-time algebraic estimators (non-asymptotic state estimation).In fact, the influence of the initial conditions is indeed removed as claimed by Fliess and Sira-Ramirez [4]. This truly is an improvement over the classical observers, which need the right initial conditions. Unknown or incorrect initial conditions invariably entail slow convergence of recursive type of observers. In addition, the presence of integrals in the estimation procedure acts like a low pass filter, which naturally reduces the influence of noise and external perturbation and hence is good at estimating the damping force of MR damper from a noisy signal. Combined with algebraic tools, in the future, a new active disturbance rejection control law can be formulated for achieving a good performance with excellent tracking accuracy and while offering insensitivity to unpredictable disturbance.

The organization of the paper is as follows: a simple car model and the description of its mathematical equations are presented in Sect. 2. Section 3 describes the algebraic estimator and its principle rules for implementation process. The effectiveness of the proposed estimator in diagnosing MR damper failure is illustrated in Sect. 4. Finally, the conclusion is given in Sect. 5.

2 Vehicle Model

A model of vehicle with two degrees of freedom is considered as the most basic model that could describe automotive suspension (Fig. 1). It consists of an assumption based on considering that the total mass of vehicle is equally distributed among the four wheels. Only vertical movements are considered. Dampers or springs prevent the amplification of disturbances caused by road profile while maintaining good road contact. The selected simplified model is helpful, for a first study, to validate the proposed estimator. Failures of damping system can be caused by various factors: unpredictable leak of fluid of MR, breakage of assembly support, loosening of joints, under-inflation or over-inflation of tires, wear of the tire etc. [2]. The dynamic behavior of a quarter-car model with a semi-active suspension is described by:

$$m_s \ddot{x}_s = -F_d - k_s(x_s - x_u) \tag{1}$$

$$m_u \ddot{x}_u = k_s(x_s - x_u) + F_d - k_t(x_u - x_r) \tag{2}$$

where, x_s, x_u and x_r are the sprung mass displacement, unsprung mass displacement and road profile excitation, respectively; the chassis is represented by m_s, the wheel and the tire are represented by m_u. k_s is the suspension stiffness and k_t is the tire stiffness. The damper in this case, called "controllable damper", is able to generate a control force for providing the adequate ride comfort. This damping force with hysteresis effect can be expressed by different ways following different laws such those given by Bingham, Bouc-Wen, LuGre and Dahl models. In this study, the plastic model, that is the simplest one known by "Bingham model", is used. It is expressed as following:

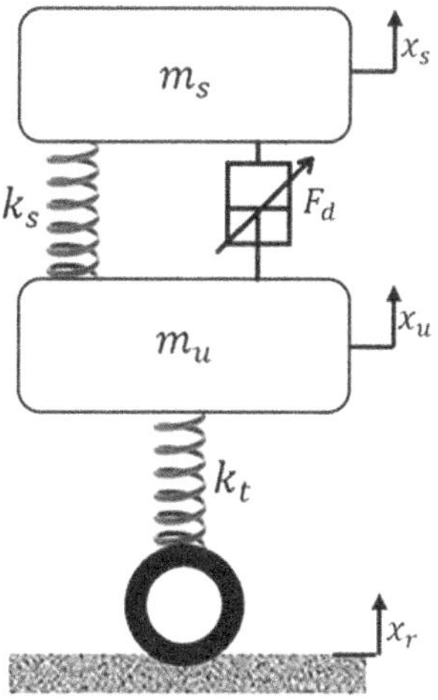

Fig. 1 Semi-active quarter car model

$$F_d = \beta\big[k_p(x_s - x_u) + d_p(\dot{x}_s - \dot{x}_u) + f_c \upsilon \tanh[a_\upsilon(\dot{x}_s - \dot{x}_u) + a_d(x_s - x_u)]\big] \quad (3)$$

The viscous damping coefficient is d_p. The stiffness coefficient is k_p.An electric variable is given by υ (control input). The hysteretic behavior is characterized by a_υ and a_d. The direction of control force depends on the change of suspension deflection velocity $(\dot{x}_s - \dot{x}_u)$ and position $(x_s - x_u)$.f_c is offset force.

3 Proposed Algebraic Estimator

As cited in Sect. 1, there are failures that can threat the operation of shock absorber. Our objective is to identify and estimate it. Non-asymptotic algebraic estimator proposed by Haddar et al. [5] for identification of road profile showed interesting effectiveness. It is selected for the present paper. The differential algebra and operational techniques are the key elements of the proposed scheme. The proposed estimator can be integrated for many semi-active control system implementations [6]. In Eq. (1), which relates damping force and vehicle dynamics response, F_d signal can be temporarily approximated by a step function ϕ [7]. Under this assumption, for a short time period, the damping force can be given as follows:

$$\phi = -k_s(x_s - x_u) - m_s\ddot{x}_s \quad (4)$$

Regarding the structure of the estimation algorithm, different steps should be followed:

Step 1. Transition from time domain to the Laplace domain:

$$\frac{\phi}{s} = -k_s(X_s(s) - X_u(s)) - m_s\big(s^2X_s(s) - sX_s(t_0) - \dot{x}_s(t_0)\big) \quad (5)$$

Step 2. Elimination of initial conditions based on double differentiation of Eq. (5):

$$2\frac{\phi}{s^3} = -k_s\left(\frac{d^2X_s(s)}{ds^2} - \frac{d^2X_u(s)}{ds^2}\right) - m_s\left(2X_s(s) + 4s\frac{dX_s(s)}{ds} + s^2\frac{d^2X_s(s)}{ds^2}\right) \quad (6)$$

Step 3. Cancelling out the positive power (it is defined by the highest degree of s) Both sides are multiplied by s^{-3}:

$$2\frac{\phi}{s^6} = -k_s\left(\frac{1}{s^3}\frac{d^2X_s(s)}{ds^2} - \frac{1}{s^3}\frac{d^2X_u(s)}{ds^2}\right) - m_s\left(2\frac{X_s(s)}{s^3} + 4\frac{1}{s^2}\frac{dX_s(s)}{ds} + \frac{1}{s}\frac{d^2X_s(s)}{ds^2}\right) \quad (7)$$

Step 4. Back to time domain.

Based on the previous steps, the algebraic estimator of damping force may be expressed in terms of verticals displacements $x_s(t)$ and $x_u(t)$ as follows:

$$\widehat{F}_d(t) = -\frac{30}{L^5}\int_0^t k_s(L-\tau)^2\tau^2(x_s(\tau) - x_u(\tau))d\tau - \frac{60}{L^5}\int_0^t m_s\left(L^2 - 6\tau L + 6\tau^2\right)x_s(\tau)d\tau \tag{8}$$

The presence of damper defect is modeled as following:

$$\widehat{F}_d(t) = \overline{F}_d(t) - F_\delta(t) \tag{9}$$

where healthy damper is presented by a nominal force $\overline{F}_d(t)$ and the loss of effectiveness due the additive perturbation (induced by sensors noise) is given by $F_\delta(t)$.

For multiplicative default, such as oil leakage fault, β can be modeled with a simple relation as following:

$$\overline{F}_d(t) = \frac{\widehat{F}_d(t)}{\beta} \tag{10}$$

The effectiveness of the damper is described by β (A healthy damper is given by $\beta = 1$ and $0 \leq \beta < 1$ describes a damper failure). For estimating β and in order to avoid the problem of singularities (for more details see: Alvarez-Sánchez [8] and Beltran-Carbajal et al. [9]), we will use $\hat{\beta}$ which is expressed as:

$$\hat{\beta} = \frac{\iint t^2\widehat{F}_d(t)}{\iint t^2\overline{F}_d(t)} \tag{11}$$

4 Results of Simulation

The quarter car parameters are presented in Table 1.

Scenario 1: *Additive noise in sprung mass sensor*

The road profile excitation is given by $x_r(t) = 0.0375\sin(2\pi(7.77)t)$. The sprung mass sensor was affected by white Gaussian noise (Fig. 2). The imposed algorithm can detect this kind of perturbation as depicted in Fig. 3. The Hysteresis behavior damping force from estimated signal has a similar behavior of real hysteresis.

Table 1 Vehicle parameters [2]

Parameters	Value
m_s	315 kg
m_u	37.5 kg
k_s	29,500 N/m
k_t	230,000 N/m
k_p	– 10,239 N/m
d_p	1500 Ns/m
f_c	441 N
υ	2 A
a_υ	7.89 s/m
a_d	– 13.8 m^{-1}

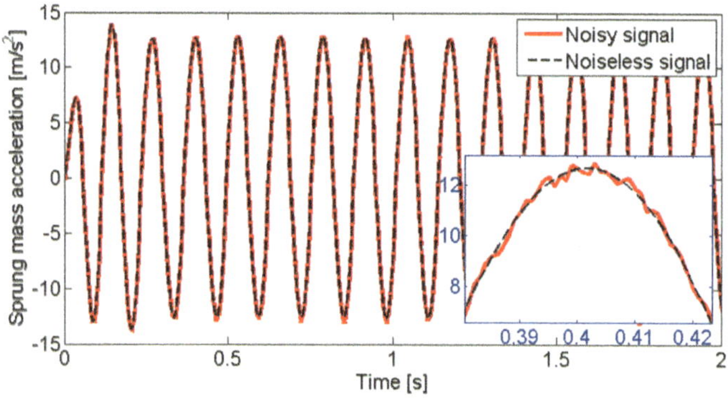

Fig. 2 Additive failure in sprung mass acceleration sensor

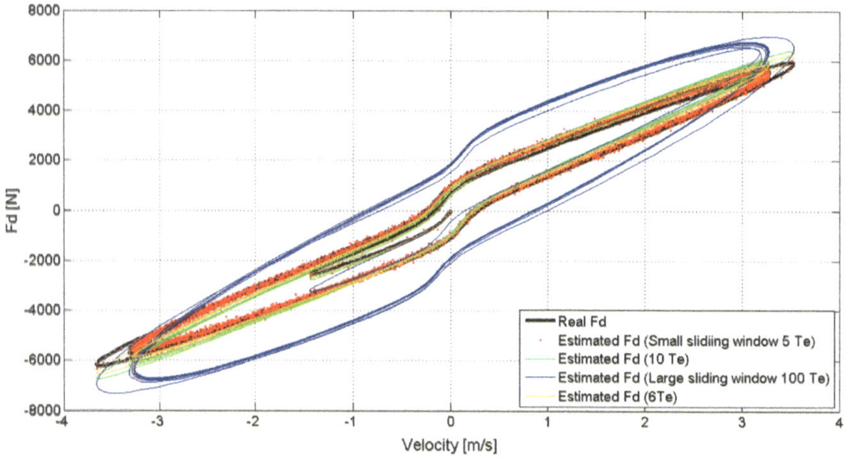

Fig. 3 Estimated Hysteresis behavior damping force with different values of L

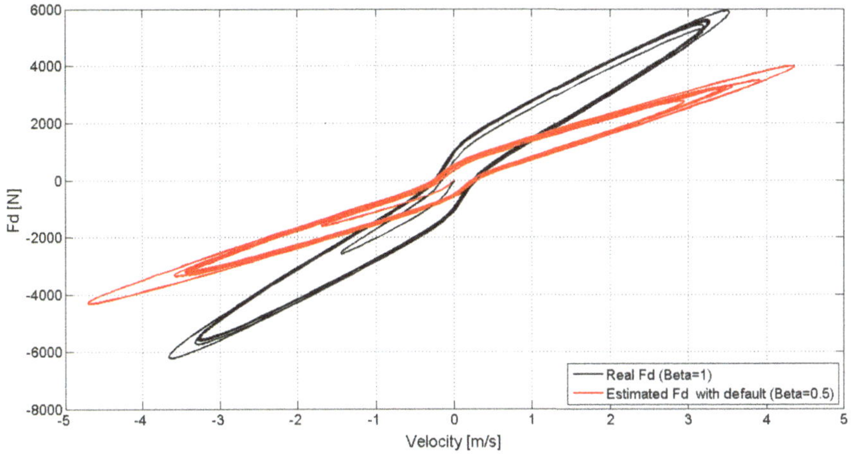

Fig. 4 Hysteresis behavior of Bingham model with multiplicative failure

However, in the case of algebraic estimator, the size of sliding windows L can affect the precision of estimation. In fact a large length of L will filter noise and a delay can appear in the simulation. This delay is observed by the largest dynamic of hysteresis loop shape (Blue curve).These information can be used in classical control scheme as PID control for getting an error value $e(\mathrm{t})$ as the difference between a desired response $\overline{F}_d(t)$ and a estimated process variable $\widehat{F}_d(\mathrm{t})$ for applying a correction based on proportional, integral, and derivative terms.

Scenario 2: *Multiplicative perturbations*

The multiplicative damper fault is modelled as a step change at t = 5 s from $\beta = 1$ (healthy case) to $\beta = 0.5$ (unhealthy case). Figure 4 shows how the multiplicative perturbation influences the dynamic of hysteresis loop shape and damping force. So, it is important to detect this kind of failure in shock absorber. Figure 5 describes a satisfactory estimation of the β coefficient after a short time $\varepsilon = 0.1s$.This kind of information can be considered as novelty detection to detect abnormal behavior in suspension system.

5 Conclusion

This work proposes an algebraic estimator of MR damping force. It is a simple scheme able to identify the additive and multiplicative failures that affects the precision in hysteresis. According to operational rules, an algebraic estimator was implemented based on the assumption that the damping force is a constant piecewise function. Hence, this work is helpful to get more information about vehicle behavior online

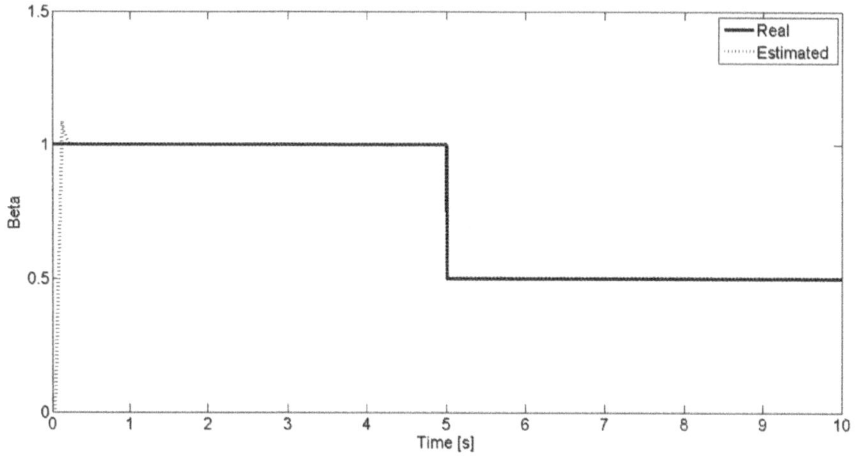

Fig. 5 Algebraic estimation of multiplicative failure

for semi-active controller evaluation purposes. The main finding can be summarized the following:

- The introduced scheme exploits only the vertical displacements of the suspension system.
- One scaling parameter is needed for setting the estimator and achieving a good fault estimation.
- Additive and multiplicative failures can be identified in the same time.

After this condition monitoring, the estimated disturbances can be rejected online to an intelligent controller that will be the subject of future work.

References

1. Zhang H, Tangirala AK, Shah SI. (1999). Dynamic process monitoring using multiscale PCA. In: Engineering solutions for the next millennium. 1999 IEEE Canadian conference on electrical and computer engineering (Cat. No. 99TH8411), vol 3, pp 1579–1584. IEEE
2. Hernandez-Alcantara D, Morales-Menendez R, Amezquita-Brooks L (2015) Fault detection for automotive shock absorber. J Phys Conf Ser 659(1):012037
3. Odendaal HM, Jones T (2014) Actuator fault detection and isolation: an optimised parity space approach. Control Eng Pract 26:222–232
4. Fliess M, Sira-Ramirez H (2003) An algebraic framework for linear identification. ESAIM Control Optimisation Calculus Vari 8:151–168
5. Haddar M, Baslamisli SC, Chaari R, Chaari F, Haddar M (2019) Road profile identification with an algebraic estimator. Proc Inst Mech Eng C J Mech Eng Sci 233(4):1139–1155
6. Eshkabilov S (2016) Modeling and simulation of non-linear and hysteresis behavior of magneto-rheological dampers in the example of quarter-car model. *arXiv preprint* arXiv:1609.07588.
7. Fliess M, Join C (2013) Model-free control. Int J Control 86(12):2228–2252

8. Alvarez-Sánchez E (2013) A quarter-car suspension system: car body mass estimator and sliding mode control. Procedia Technol 7:208–214
9. Beltran-Carbajal F, Silva-Navarro G (2017) A fast parametric estimation approach of signals with multiple frequency harmonics. Electric Power Syst Res 144:157–162

On the Use of Jerk for Condition Monitoring of Gearboxes in Non-stationary Operations

Fakher Chaari, Stephan Schmidt, Ahmed Hammami, P. Stephan Heyns, and Mohamed Haddar

Abstract Diagnostics of rotating machinery is very important to preserve their efficiency. Defects should be detected at an early stage in order to plan maintenance and avoid a stop in production. In this chapter the use of jerk, the derivative of acceleration, is used to diagnose a gearbox with a local tooth defect. A dynamic gearbox model is presented, which includes time-varying mesh stiffness and non-stationary operating conditions modeled as variations of load and speed. In order to model the tooth defect, a reduction in mesh stiffness is introduced proportionally to the severity of the defect. Two case studies are presented. The first one concerns stationary operating conditions and different severity of defects. For this case, the jerk shows good ability to diagnose the presence of the local defect. For the second case, variable loading condition was modeled as a sawtooth shape, whereafter the jerk was used to detect the presence of a defect. The amplitude modulation cause in increase in the vibration level that does not allow the identification of defect using only the acceleration signal. The jerk allows this identification even for significant load variability. The performance of jerk for signals with additive Gaussian noise is discussed, highlighting its limitations.

Keywords Jerk · Diagnostic · Gear system · Defect · Modelling

1 Introduction

Gearboxes play an important role in power transmission from a driving unit to a receiver. They are characterized by their high efficiency and robustness. However, overloads, variable loads and speeds can be a serious threat for their good operation causing harmful damages. Condition monitoring of such gearboxes using vibration analysis becomes more difficult and requires advances signal processing techniques.

F. Chaari (✉) · A. Hammami · M. Haddar
Laboratory of Mechanics Modelling and Production, National School of Engineers of Sfax, BP 1173, 3038 Sfax, Tunisia

S. Schmidt · P. S. Heyns
Centre for Asset Integrity Management, University of Pretoria, Pretoria, South Africa

F. Chaari et al. (eds.), *Smart Monitoring of Rotating Machinery for Industry 4.0*, Applied Condition Monitoring 19, https://doi.org/10.1007/978-3-030-79519-1_10

Many papers discussed such techniques and highlighted the modulation effects that appear when both load and speed conditions vary. Chaari et al. [1] proposed a dynamic model of a single stage gearbox running under time varying loading conditions. The authors correlate speed and load and showed the simultaneous amplitude and frequency modulation. Amplitude modulation is induced by load variation and frequency modulation is caused by speed variation. Other works were interested in the modulation sidebands [2] that are present in the vibration spectra of planetary gearboxes.

When a local defect affects one tooth of transmission, and in the presence of non-stationary operating conditions, the monitoring task will be more complicated. Time frequency analysis is one of the most commonly used signal processing techniques that can be used in such situation [3]. But when the modulation effect is dominant the vibration signature of the defect will be lost, which impedes the detectability of the damage.

In recent years, jerk, which is defined as the first derivative of acceleration, drew the attention of the scientific community in the field of condition monitoring of rotating machinery. This was first observed for bearing diagnostics. One of the first works done was by Smith [4] who compared vibration responses obtained from shock pulse, acoustic emission and jerk measurements emitted from a damaged bearing with different levels of speed. He concluded that jerk is more efficient in detecting bearing defects than the other sensors, with the jerk improving the signal-to-noise ratios of the damage. However, it is more sensitive to the rate of rise of acceleration. Zhang et al. [5] implemented data mining algorithms and statistical methods to investigate jerk vibration signal measured on a wind turbine gearbox. They were able to detect defects in the intermediate- and high-speed stages of the gearbox which, however, was driven in stationary conditions. Ismail and Klausen [6] used jerk to localise and quantify bearing defects through the use of an autonomous fault detection technique that also includes multiple defects.

Ismail et al. [7] investigated bearing defect severity by proposing a jerk energy gradient which was applied on the synchronous average of fault impact signal. This method was successfully applied for bearings operating under low speeds. Ismail et al. [8] tested an automated vibration-based technique to predict the size of a spall defect in a bearing. This was performed by extracting from the jerk time instants. When the ball enters and exits the spall. The aim of this chapter is to confirm the ability of jerk to monitor the health status of a gearbox in the presence of both non-stationary operations and local tooth defect.

A dynamic model of a bevel gear will be presented allowing the consideration of variable operating conditions and local defect. A parametric study will be performed to investigate the efficiency of jerk for diagnostic purposes.

2 Dynamic Model

A single spiral bevel gearbox is studied. Theoretical background for the modelling procedure of this gearbox can be found in [9]. The model includes 10 degrees of freedom (6 translational and 4 rotational). The main excitation source of the gearbox is the time varying mesh stiffness which characterize the time succession 1 pair/2 pairs of teeth in contact. The geometry of the gearbox causes the shape of this mesh frequency to be trapezoidal as shown in Fig. 1. In the presence of a local defect which is chosen to be a crack, a periodic reduction of the mesh stiffness is operated proportionally to the severity of the defect. Since the transmission is driven by an asynchronous motor, load and speed are intimately related. Increase of load causes reduction of motor speed and vice versa. This will affect the periodicity of the mesh stiffness function as shown in Fig. 1 [10]

The differential equation of motion is solved using the Newmark algorithm. This algorithm provides displacement, velocity and acceleration for each time instant for the different degrees of freedom.

The jerk expresses the rate of change of the acceleration and is considered as a good indicator of vibration impulses caused by damage. For acceleration computed for continuous time, the jerk can be expressed by:

$$J = \frac{da}{dt}$$

where a is the acceleration and t is time. When the difference between two consecutive time instants become small, this expression can be expressed by:

$$J = \frac{a_{i+1} - a_i}{t_{i+1} - t_i} = \frac{a_{i+1} - a_i}{T_s}$$

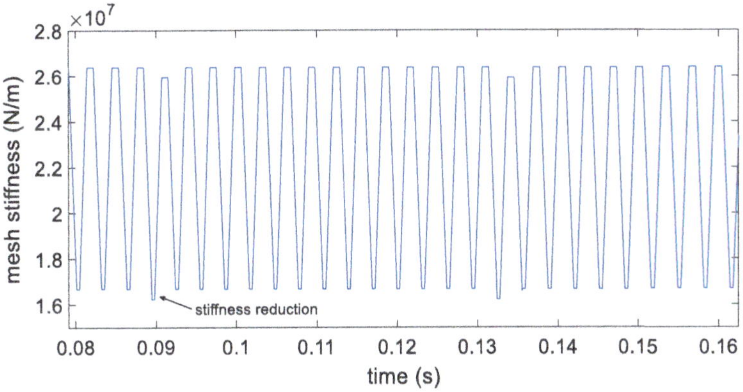

Fig. 1 Mesh stiffness evolution in case of local defect

where a_i and a_{i+1} are two successive computed accelerations for the two time instants t_i and t_{i+1} time instants. T_s is the sampling time.

3 Numerical Simulations

For stationary conditions, where speed and load are constant, the mesh frequency fm is equal to 308 Hz. A sawtooth loading condition as applied to the output of the transmission with a frequency of 5 Hz is shown in Fig. 2. Three loading amplitude rates of increase are considered: 0, 10, 25 and 50%. For the local defect different sizes of defect are considered. They correspond to the 0% (healthy case), 1, and 5%, decrease in the mesh stiffness function corresponding to increased severity of defect. In all presented simulations, comparisons will be made between acceleration response and jerk calculated using the response of the pinion bearing.

3.1 Stationary Operating Conditions

In this section, the dynamic response of the transmission is simulated considering constant loading conditions. Figure 3 shows the acceleration for different size defects.

From the simulated acceleration, it is well noticed that acceleration allows the detection of the defect starting from 5% severity. Peaks shown for this level confirm clearly the presence of defects with visible periodic behavior since speed is constant. However, for the case of incipient defect (1%), the acceleration signal does not allow us to detect the presence of the defect. Figure 4 shows the jerk for the cases with 1 and 5% defect sizes.

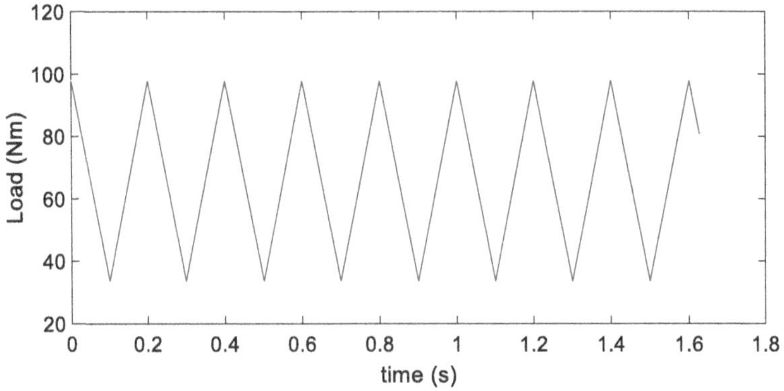

Fig. 2 Load fluctuation

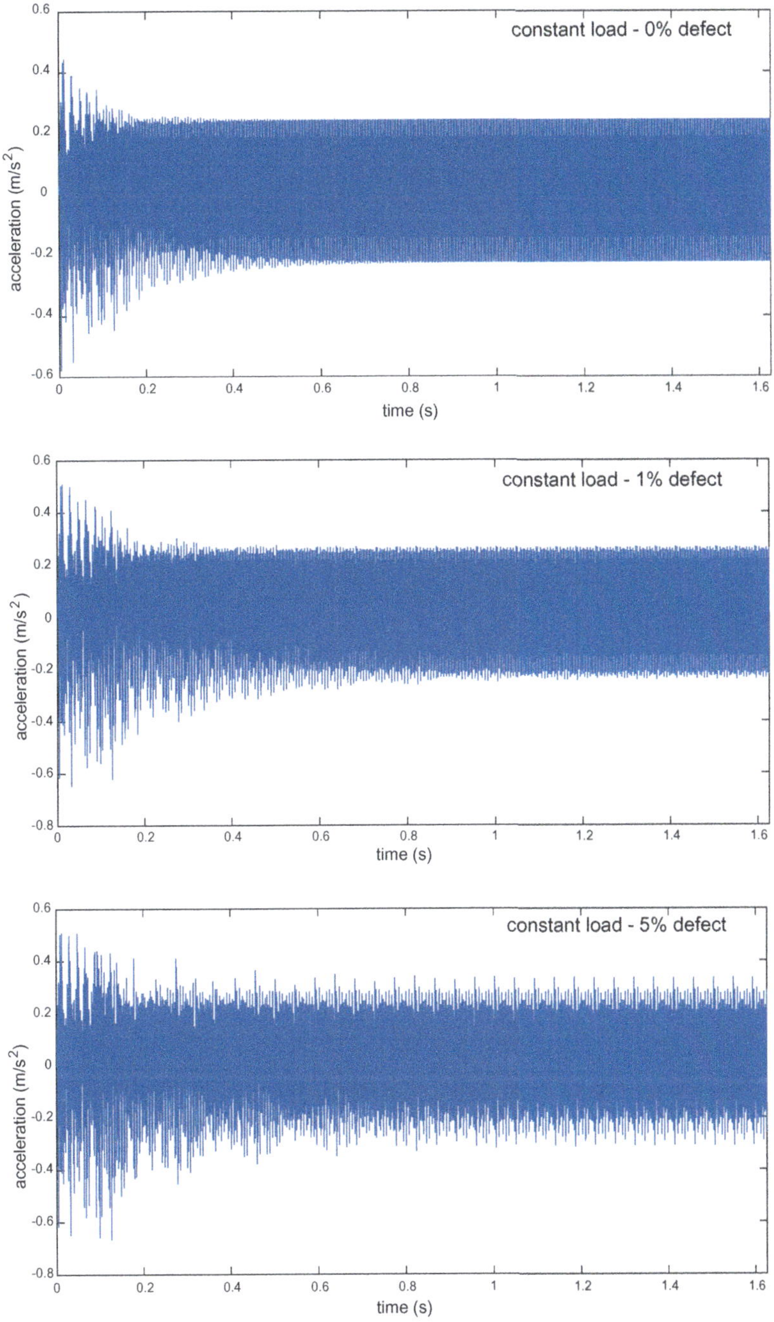

Fig. 3 Acceleration for constant load and 2 levels of defect

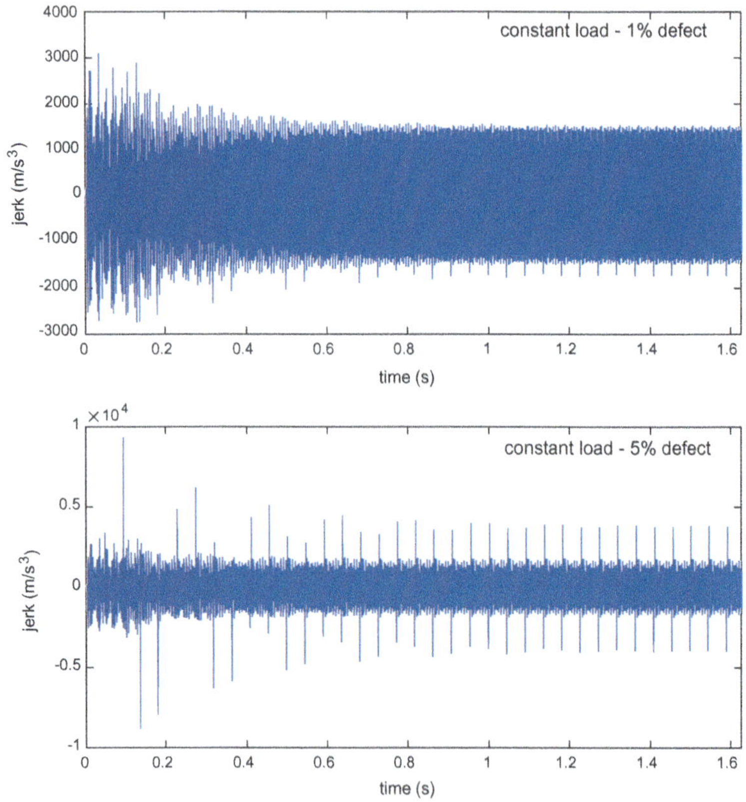

Fig. 4 Jerk for constant load and two levels of defects

It is well noticed that jerk allows the identification of the defect since its incipient form. For 1% defect severity spikes relative to the defect are well identified and they have higher amplitudes especially for the case of 5% severity.

3.2 Non-stationary Operating Conditions

Now the transmission is subjected to the sawtooth varying load as specified in Fig. 2. We will focus on 5% defect size case. Figure 5 show the acceleration simulated for 10, 25 and 50% increase of the load.

From the acceleration signals, it is not possible to find a clear impulse to confirm the presence of the defect. The amplitude modulation induced by the load variation dominates the dynamic response. Figure 5 shows the jerk computed for the last acceleration response (Fig. 6).

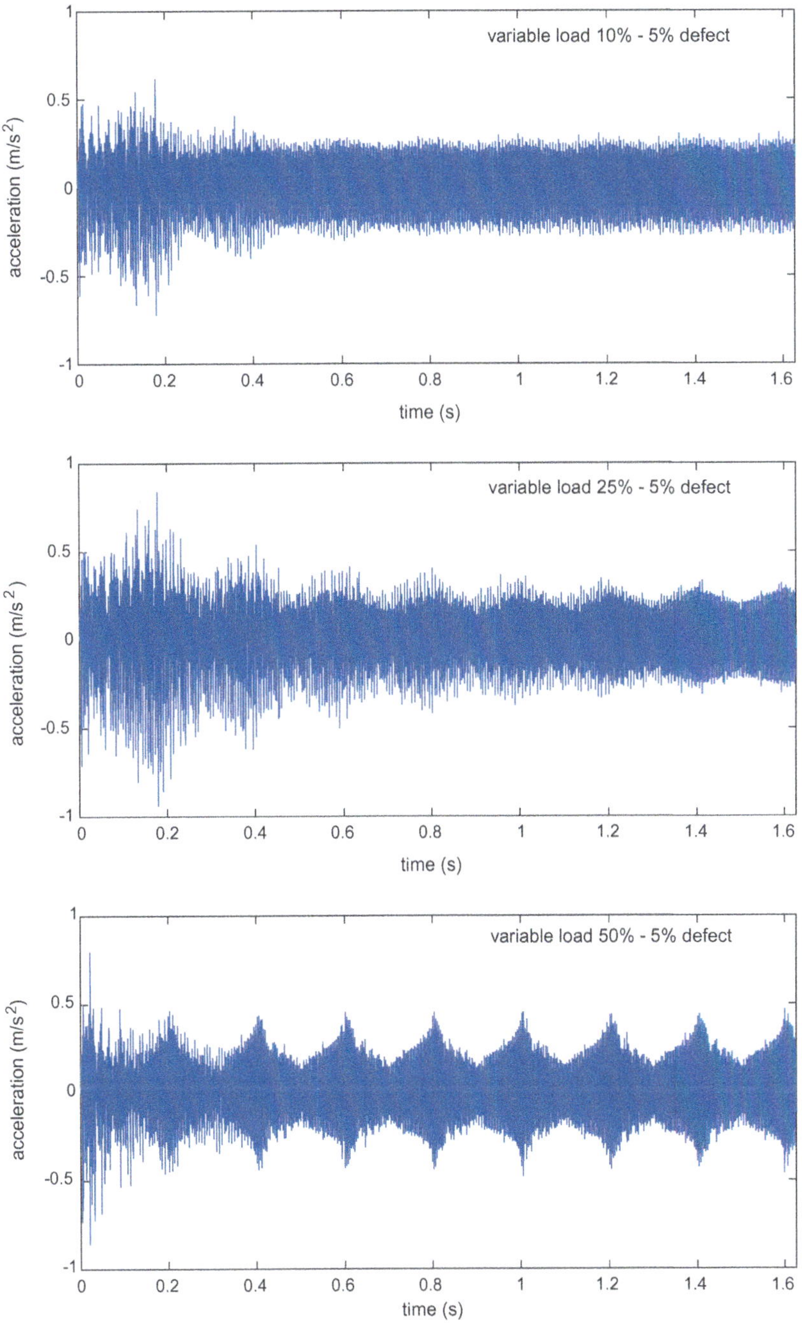

Fig. 5 Acceleration for 5% severity defect and different loading conditions

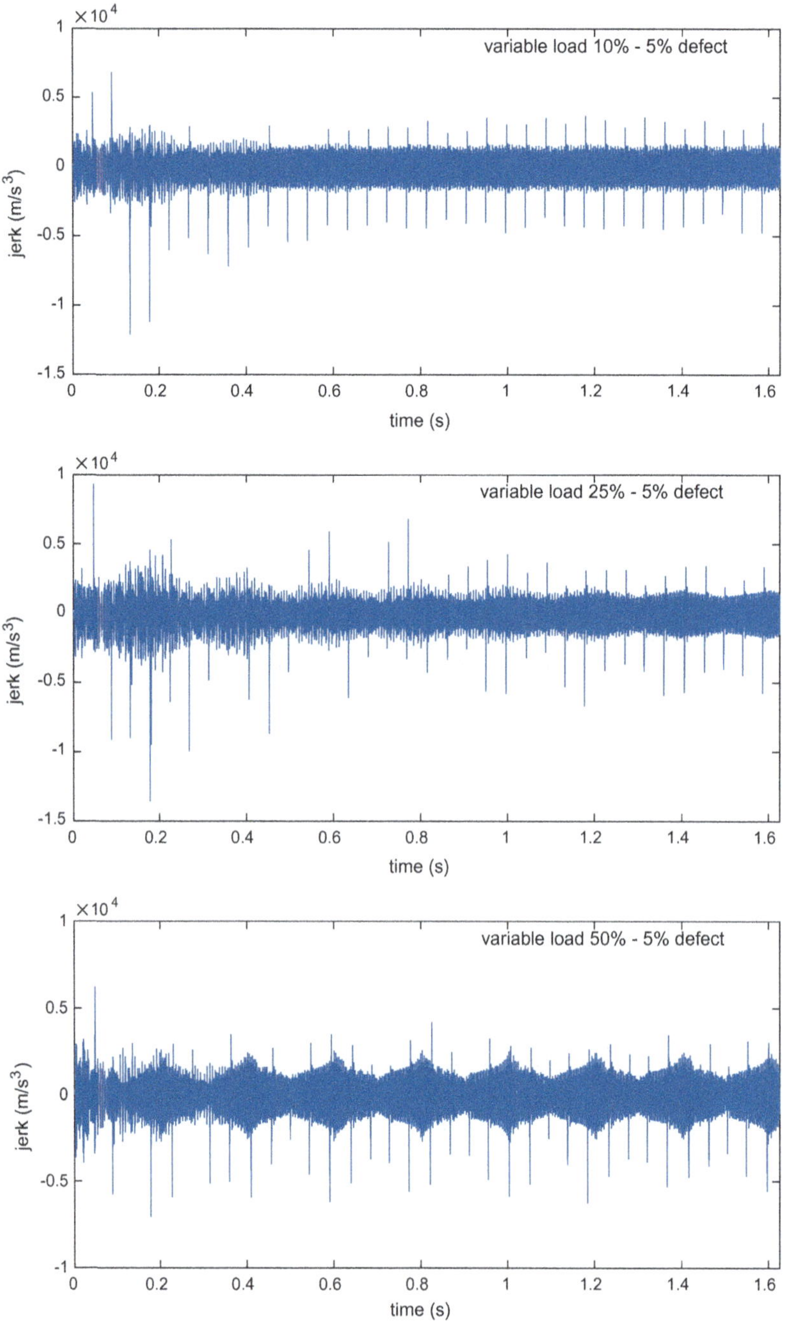

Fig. 6 Jerk evolution for 5% severity defect and different loading conditions

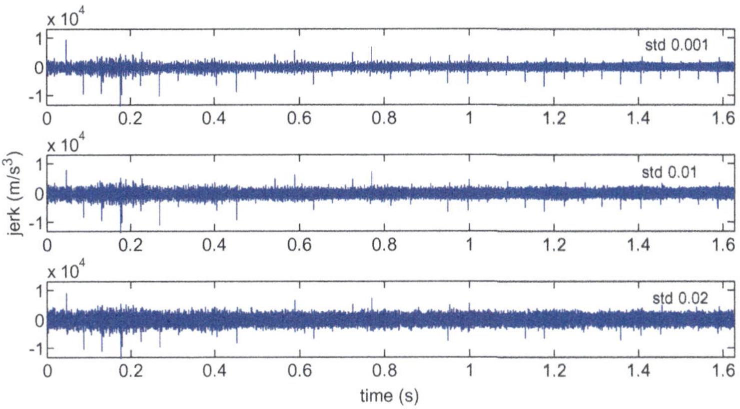

Fig. 7 Jerk evolution for 5% defect severity, 25% load variation and three cases of standard deviation

It is clear that jerk allows the identification of defect very early. The amplitude of the spikes is varying with high amplitude when the defected tooth mesh and the load is maximum and vice versa.

3.3 *Influence of Noise*

In this part, the robustness of using jerk as a condition monitoring parameter when noise is present in the signal is investigated. White Gaussian noise is used with three cases of standard deviations which are 0.001, 0.01, 0.02 and 0.05. The noise is added to simulated acceleration for the case 25% load variation and 5% defect severity. Figure 7 shows the jerk plotted for the three cited cases.

It is noticed that jerk is sensitive to the increase of noise. If the value of 0.02 in standard deviation is exceeded, the efficiency of detection is lost. This is attributed to the central difference operator used in the estimation of the jerk from the measured vibration signal. For example, the subtraction of two zero mean Gaussian random variables results in a new zero mean Gaussian variable with a variance larger than the individual Gaussian components, i.e. the variances are additive. This fact was highlighted by [6] who stated that due to strong background noise, spikes caused by faulty bearings cannot be identified accurately in the raw vibration jerk signal.

4 Conclusion

The objective of this chapter is to check the efficiency of using jerk as a condition monitoring tool to identify the presence of local defects in gear systems. To this effect, a numerical model of a bevel gear transmission was proposed. Several loading conditions were studied. A local tooth defect was modeled as a reduction in the mesh stiffness function. A parametric study was conducted by combining three cases of loading conditions and two cases of defect severity.

Accelerations were simulated and compared to jerk. Some interesting findings can be summarized as follows:

- For constant loading conditions and by checking acceleration signals, it is difficult to identify the presence of defect in its incipient stage. However, the jerk time signal allows the identification of the defect at a very early stage.
- For time-varying loading conditions, the acceleration signal is unable to reveal the presence of the defect because of the important amplitude modulation caused by the load increase. Jerk is sensitive to the presence of the defect starting from very small severity of defect and for high fluctuation of load.

However, presence of noise can alter the identification of defects. It is recommended to de-noise the signal. It is interesting to investigate more the implementation of jerk in signal processing by checking more case studies and implementing more deep frequency and time–frequency analysis.

Acknowledgements The South African and Tunisian authors acknowledge the South African and Tunisia Research Cooperation Programme 2019 (SATN 180718350459) for partially supporting this research.

References

1. Chaari F, Bartelmus W, Zimroz R, Fakhfakh T, Haddar M (2012) Gearbox vibration signal amplitude and frequency modulation. J Shock Vibr 19:635–652
2. Inalpolat M, Kahraman A (2009) A theoretical and experimental investigation of modulation sidebands of planetary gear sets. J Sound Vib 323(3–5):677–696
3. Bartelmus W, Chaari F, Zimroz R, Haddar M (2010) Modelling of gearbox dynamics under time-varying nonstationary load for distributed fault detection and diagnosis. Eur J Mech A/Solids 29(4):637–646
4. Smith JD (1982) Vibration monitoring of bearings at low speeds. Tribol Int 15(3):139–144
5. Zhang Z, Verma A (2012) Fault analysis and condition monitoring of the wind turbine gearbox. IEEE Trans Energy Convers 27(2)
6. Ismail MAA, Klausen A (2018) Multiple defect size estimation of rolling bearings using autonomous diagnosis and vibrational jerk. In: The 7th World conference on structural control and monitoring (7WCSCM), Qingdao, China, 22–25 July 2018
7. Ismail MAA, Sawalhi N, Pham T (2015) Quantifying bearing fault severity using time synchronous averaging jerk energy. In: ICSV22, Florence (Italy), 12–16 July 2015
8. Ismail MAA, Bierig A, Sawalhi N (2017) Automated vibration-based fault size estimation for ball bearings using Savitzky–Golay differentiators. J Vibr Control 24(18). https://doi.org/10.1177/1077546317723227

9. Karray M, Feki N, Khabou MT, Chaari F, Haddar M (2017) Modal analysis of gearbox transmission system in Bucket wheel excavator. J Theor Appl Mech 55(1):253–264
10. Karray M, Chaari F, Khabou MT, Haddar M (2017) Dynamic analysis of bevel gear in presence of local damage in nonstationary operating conditions. In: Haddar M, Chaari F, Benamara A, Chouchane M, Karra C, Aifaoui N (eds) Design and modeling of mechanical systems—III. CMSM 2017. Lecture notes in mechanical engineering. Springer, Cham. https://doi.org/10.1007/978-3-319-66697-6_32

Dynamic Remaining Useful Life Estimation for a Shaft Bearings System

Mohamed Habib Farhat, Fakher Chaari, Xavier Chiementin, Fabrice Bolaers, and Mohamed Haddar

Abstract Condition-based maintenance of rotating machines has become a subject of growing interest in 4.0 industry. Significant failures in industrial equipment are directly related to bearing degradation. Many techniques have been used successfully for diagnosing and forecasting bearing failures. However, accurately estimating the remaining useful life (RUL) of a bearing in operation remains a challenge. In industrial applications, the difficulty is usually related to the lack of available historical degradation signals. Therefore, in this chapter, a prognostic approach addressing the shortcomings of historical degradation data is presented. The latter bases on real-time acquired signals to build an adaptive predictive model for bearing degradation. Initially, diagnostic features associated to bearings are extracted from the available vibration signals. Then, an unsupervised DBSCAN classifier is used to detect degradation. As new degradation data become available, the extracted features are interactively ranked according to a defined selection criterion. The relevant feature is chosen as health indicator (HI). The degradation evolution and the RUL are estimated by an applied adaptive exponential degradation model, dynamically updated with each acquired sample. The applied strategy has been validated against vibration data acquired from a real wind turbine's shaft bearings system.

Keywords Bearing · Prognostic · RUL · Degradation

1 Introduction

Following the 4th industrial revolution, conventional maintenance strategies (conditional, preventive) have gradually given way to a more futuristic and intelligent vision of maintenance, namely: the predictive maintenance. This approach allows real-time

M. H. Farhat (✉) · F. Chaari · M. Haddar
Laboratory of Mechanics, Modeling and Production (LA2MP), National School of Engineers of Sfax, BP1173, 3038 Sfax, Tunisia
e-mail: Mohamed-habib.farhat@enis.tn

M. H. Farhat · X. Chiementin · F. Bolaers
Institute of Thermics, Mechanics and Material (ITHEMM), University of Reims, Moulin de La Housse, 51687 Reims cedex 2, France

F. Chaari et al. (eds.), *Smart Monitoring of Rotating Machinery for Industry 4.0*, Applied Condition Monitoring 19, https://doi.org/10.1007/978-3-030-79519-1_11

monitoring of failures and provides reliable RUL prediction [1]. Bearings are recognized among the most critical components of industrial machinery. An uncontrolled degradation in its state could be the cause of catastrophic damages. Correspondingly, tracking bearing failures and forecasting its RUL have become a subject of significant interest. In the literature, RUL prediction methods are divided into two categories: data-based methods and model-based methods [2]. Data-based approaches derive machines degradation processes based on the available measurement signals without the need to model the actual physical equipment (case of the model-based approach). These latter are based on the assumption of relatively consistent statistical characteristics in machines vibration in case of normal operation conditions. Data-driven method has proven to be an increasingly promising approach for bearing's monitoring and RUL prediction [3]. The degradation trend is first identified by analyzing the condition-monitoring signals. Then, future evolutions are anticipated using a prediction model. Unsupervised machine learning methods have proven to be effective in the detection of bearing defects [4]. Unlike supervised classifiers, the use of unsupervised clustering algorithms requires no prior knowledge about the operation data. Among density-based clustering algorithms, Kerroumi et al. [5] tested and validated the efficiency of DBSCAN in the diagnosing of bearing vibration signals. The RUL estimation methods are used to assess the equipment's degradation and to predict impending failure. To do so, an appropriate degradation feature must be carefully selected. There are many conventional features associated to bearing degradation, including time domain [6], frequency domain [7], and time–frequency domain [8]. Subsequently, an optimal prediction model must be developed to forecast the evolution of the HI through time. Many data-based prediction methods were proposed in the literature, namely, the artificial neural network [9], the adaptive neural fuzzy interference [10], the support vector machine and relevance vector machine (RVM) [11]. Nevertheless, exponential regression models remain one the simplest and efficient techniques used for bearing (RUL) estimation [12]. In most of the RUL prediction work proposed in the literature, the HI is selected based on the run-to-failure data. However, in real-world applications, these data are generally not available. In this work, a preventive defect detection approach is investigated to deal with the lack of bearings degradation data. Real run to failure shaft bearings system dataset are used to validate the proposed approach. 13 associated bearing diagnostic features are extracted from the considered vibration data. Kernel principal component KPCA [13] is used to reduce the extracted features dimension. The resulting principal components are used to feed a DBSCAN classification model, used to defect the defects. The features extracted from healthy data are arranged according to a selection criterion and the selected feature is chosen as the HI. As new degradation data become available, the extracted features are re-ranked according to the selection criterion. Each time, the selected feature is chosen as HI, which is used to build an adaptive exponential degradation model, allowing the estimation of the RUL.

The paper is organized as following. The proposed maintenance methodology is detailed in Sect. 2. The validation of the proposed approach is carried out in Sect. 3. Conclusions are performed in Sect. 4.

2 Methodology

RUL refers to the residual lifetime of the machine before it loses the ability to operate based on its current state and its past operating history. The RUL is generaly checken through a conditional random variable specified in relative or absolute time units. This variable is the HI. The machine is said to be at the end of its life if the HI reaches a limit value called the failure threshold.

Figure 1 present the predictive maintenance strategy proposed in this work for a ball bearing system. The latter is composed of two main phases: Defects detection and RUL prediction.

In order to solve the problem of limited historical data, the proposed approach is designed to perform with real-time acquired data. It is assumed to be implemented when the machine is in a healthy state. For both **phase 1** and **phase 2**, each new acquired signal is subjected to a features extraction step. This aims to identify the relevant characteristics of the vibration signal that are sensitive to system degradation.

During **phase 1** (deterioration detection phase), KPCA is first applied to all the extracted characteristics in order to extract the optimal combination that best presents the signal through the elimination of redundancy. The chosen combination is used as input to the DBSCAN classifier. The choice of this classifier is motivated by its ability to be used without prior knowledge about historical signals, dealing with historical data limitation problems. The latter makes a decision about the new acquired data by checking if it's corresponding to a healthy or a degradation state. For more detail

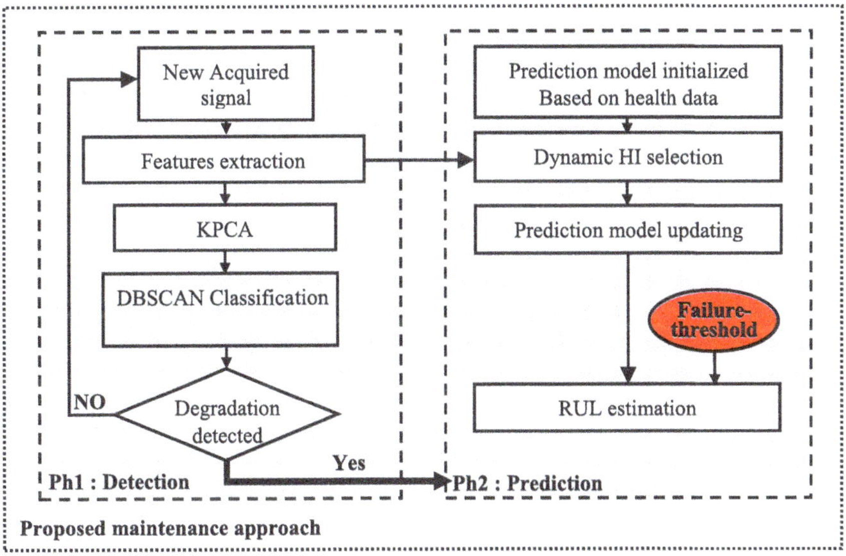

Fig. 1 Details of the proposed maintenance approach

about DBSCAN see the work of Kerroumi et al. [5]. While no defect is detected, the previous steps will be repeated for each new acquired data.

Phase 2 (RUL prediction) triggers directly once a degradation is detected. The prediction of the HI evolution will be assured in this work by an exponential prediction model defined initially based on the available healthy data. The choice of an exponential model for the RUL prediction is made according to the work of Si et al. [16], which assumes an exponential evolution for bearing degradation. The HI is selected interactively with each new acquired degradation signal. Based on the available data, a selection criterion is used to make a decision about most relevant HI.

The selection criterion adopted in this work consists in checking two main caracteristic of the features, namely: motonicity and trendability. These two metrics are confined in the range [0–1], subject to change according to the available observations and are positively correlated with the performance of the features.Therefore they are suitable to be used for the HI selection. For more detail about these two caracterestics, see [14]. The selection criterion adopted consists in summing the motonocity and trendability scores of each feature. The feature obtaining the highest score is chosen as HI:

$$S_{\text{Criterion}} = \text{Motonocity} + \text{Trendability}$$

The parameters of the prediction model are progressively rectifying as more data become available. The RUL is estimated each time by the active model according to a pre-defined failure threshold. The latter is generaly estimated based on some available failure data or it can also be set by an expert in the field. In this work, the features from the last measurement sample are taken as failure threshold. All the features are normalized between [0–1], where 1 correspondant to the failure threshold.

3 Validation of the Proposed Approach

This part studies the effectiveness of the proposed approach in estimating the RUL of a bearing. All the steps explained in the previous section are performed on a real bearing degradation data.

3.1 Experimental Setup

Referring to [15], the degradation data considered in this work are acquired from a shaft bearing system. The latter is implemented in a commercial wind turbine supplied by Green Power Monitoring Systems (GPMS) in the United States. This data is used to verify the proposed method for bearing signals diagnosis and RUL

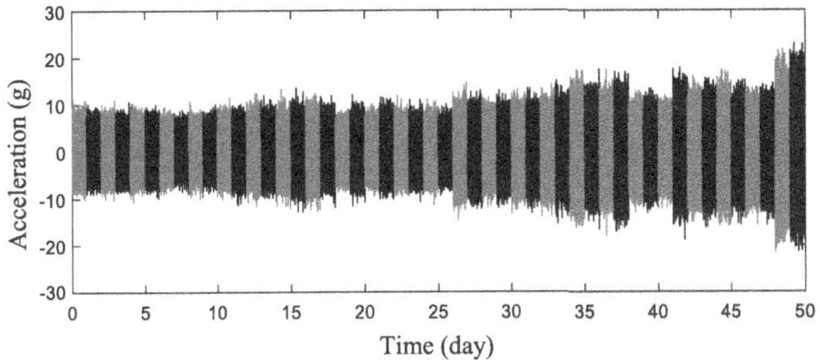

Fig. 2 Collected vibration signals

prediction. It includes a set of run to failure data. The vibration signals are recorded every day at a sampling rate of 97,656 Hz. The duration of each recorded signal is 6 s. 50 measurements are considered in total.

The raw degradation signals of the considered shaft bearing are shown in Fig. 2. In accordance with the hypothesis of Si et al.[16], the degradation of the shaft bearing is reflected by an exponential shape vibration amplitude evolution.

3.2 Results and Discussion

To prove the effectiveness of the proposed maintenance approach, the steps detailed in Sect. 2 are carried out for the shaft bearing system under consideration.

3.2.1 Features Extraction

Referring to their efficiency in diagnosing bearing vibration signals, 13-time domain features are chosen to be extracted from each degradation sample. The expressions of these features are given in Table 1. Figure 3 shows the evolution of some of the extracted features respectively with the degradation of the bearing.

It is clear from Fig. 3 that among the extracted features, some are more in line with the deterioration trend. Indeed, depending on their expression, some features are more sensitive to degradation, confirming the need for the HI selection step. As explained in Sect. 2, DBSCAN classifier is used to detect degradation and trigger the prognosis process. Referring to [5], the DBSCAN imputation parameters Eps and $Minpts$ are set to 10 and 4 respectively.

The result of the DBSCAN classification is shown in Fig. 4, where the green points correspond to healthy measurements and the red points to the degradation data. Based on DBSCAN the first 10 measurements are considered as healthy.

Table 1 The computed features, x represents the digitized signal, x_i is the sample number. $i \in (1,2\ldots N)$. RMS_0 represents the RMS value of the fault-free system that is recorded at the start of the vibration monitoring

Root mean square $\mathrm{RMS} = \left(\frac{1}{N}\sum_{i=1}^{N} x_i^2\right)^{\frac{1}{2}}$	Mean $\mathrm{mean} = \frac{1}{N}\sum_{i=1}^{N} x_i$	Energy $E = \sum_{i=1}^{N} x_i^2$
Kurtosis $\mathrm{Ku} = \frac{1}{N}\sum_{i=1}^{N}\left(\frac{x_i-\bar{x}}{\sigma}\right)^4$	Peak to Peak $\mathrm{ptp} = \max(x) - \min(x)$	Talaf $\mathrm{Talaf} = \log\left(\mathrm{Ku} + \frac{\mathrm{RMS}}{\mathrm{RMS_0}}\right)$
Peak $\mathrm{Peak} = \max(\lvert x_i \rvert)$	Standard Deviation $\mathrm{Std} = \sqrt{\frac{1}{N}\sum_{i=1}^{N}(x_i - \bar{x})}$	Thikat $\mathrm{Log}\left(\mathrm{Ku}^{\mathrm{CF}} + \left(\frac{\mathrm{RMS}}{\mathrm{RMS_0}}\right)^{\mathrm{peak}}\right)$
Crest factor $\mathrm{CF} = \frac{\mathrm{Peak}}{\mathrm{RMS}}$	Shape factor $\mathrm{Sf} = \frac{\left(\frac{1}{N}\sum_{i=1}^{N} x_i^2\right)^{\frac{1}{2}}}{\frac{1}{N}\sum_{i=1}^{N}\lvert x_i \rvert}$	
Skewness $\mathrm{SK} = \frac{1}{N}\sum_{i=1}^{N}\frac{(x_i-\bar{x})^3}{(N-1)\sigma^3}$	Impulse factor $IF = \frac{\max(x)}{\frac{1}{N}\sum_{i=1}^{N}\lvert x_i \rvert}$	

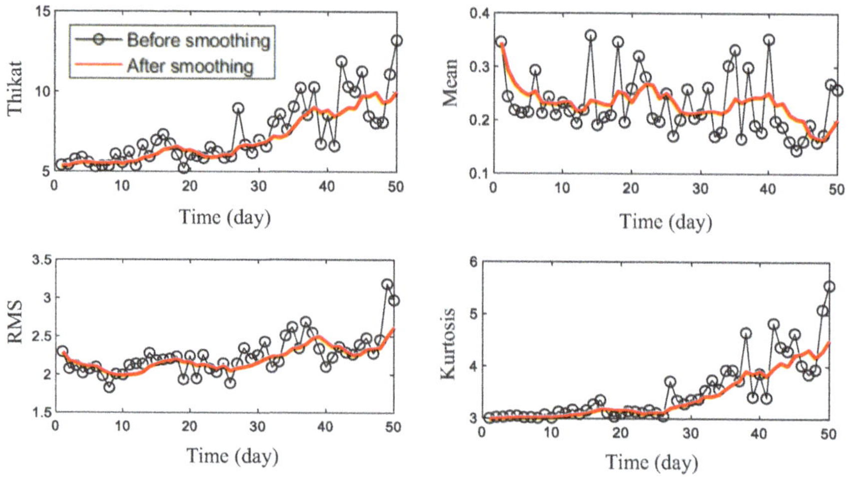

Fig. 3 Some computed features evolution (before and after smoothing)

3.2.2 RUL Estimation

Refering to [16], an exponential model is adopted to forecast the HI evolution in the shaft bearing system. The exponential degradation model is defined as:

$$\mathrm{HI}(i) = \phi + \theta \exp(\beta\, i + \epsilon)$$

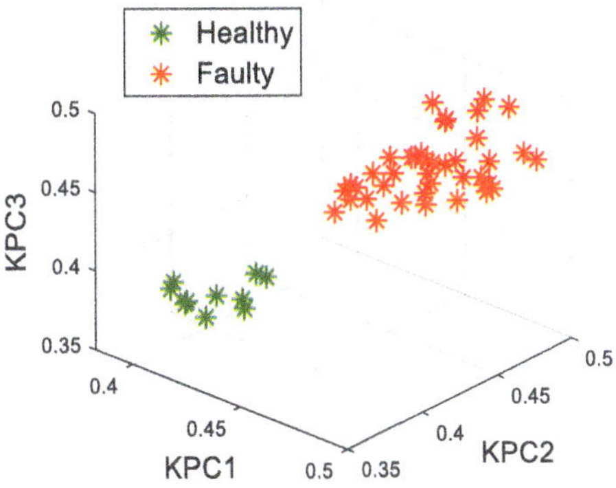

Fig. 4 DBSCAN classification result. KPC: Kernel Principal Component

HI(i) is the value of the HIcorreponding to the measurement i (day i). ϕ is an intercept term considered to be constant. θ and β are the parameters that determine the slope of the model, θ is supposed to be lognormal-distributed and β is Gaussian-distributed. ϵ is a Gaussian white noise. The parameter of the model are initialised according to the healthy data and are thenupdated interactively according to the available degradation data.

Refering to Sect. 3.2.1, the first 10 samples (measurement) of the data are related to a healthy state operation. The RUL prediction will thus be triggered from measurement 11.

As explained in Sect. 2 the HI is chosen interactively with the available data. Table 2 gives the characteristics with the highest score that are considered as a HI according to the selection criteria during the degradation days of the bearing shaft system under consideration.

According to Table 2, the features Std, Skewness, Mean and Kurtosis are used as HI to ensure the RUL estimation progressively as the system degrades. The RUL estimation is performed using the defined prediction model. In order to prove performance of the proposed dynamic HI selection method, the model is applied in two ways. Firstly, using static health indicators. Here, Fig. 5 shows the actual and the constructed model estimated RUL when using the 4 best features (selected using all degradation data) respectively as a HI, Table 3. Subsequently, the dynamic selection approach of the HI is applied. In this case, Fig. 6 shows the real RUL versus the RUL estimated using dynamic HI selection method. The RUL is given as the difference in time between the theoritical time of failure fixed in function of the chosen failure threshhold and the present time t_i, (day i).

A simple examination of Figs. 5 and 6 confirms the effectiveness of HI's dynamic selection method for estimating the system RUL. Indeed, according to Fig. 6, the

Table 2 Selected HI

Chosen HI	Std	Skewness	Mean	Kurtosis
Day	11–18	19–20	21–22	23–50

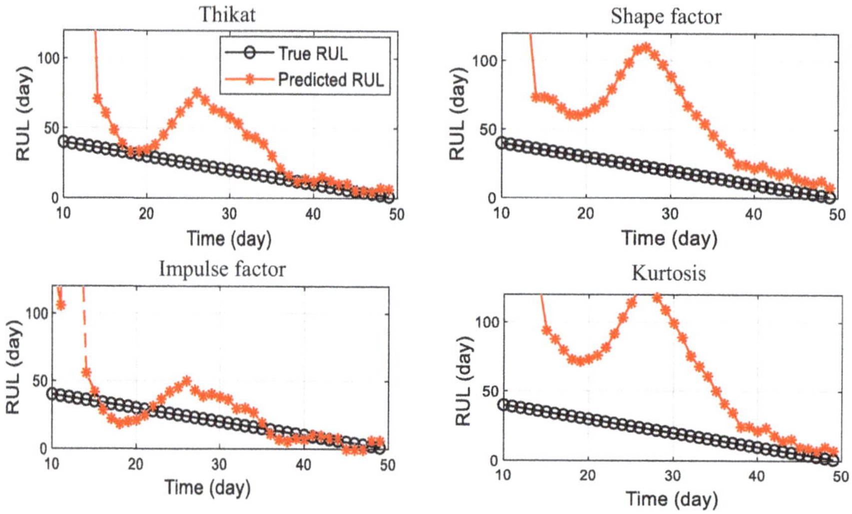

Fig. 5 Predicted RUL based on static HI

Table 3 Selection rank of the features with all degradation data available

Rank	Feature	Selection score	Rank	Feature	Selection score
1	Kurtosis	1.765	8	RMS	0.743
2	Shape factor	1.631	9	Energy	0.740
3	Impulse factor	1.311	10	Std	0.729
4	Thikat	1.175	11	Peak to peak	0.725
5	Crest factor	1.071	12	Skewness	0.607
6	Talaf	0.997	13	Peak	0.482
7	Mean	0.770			

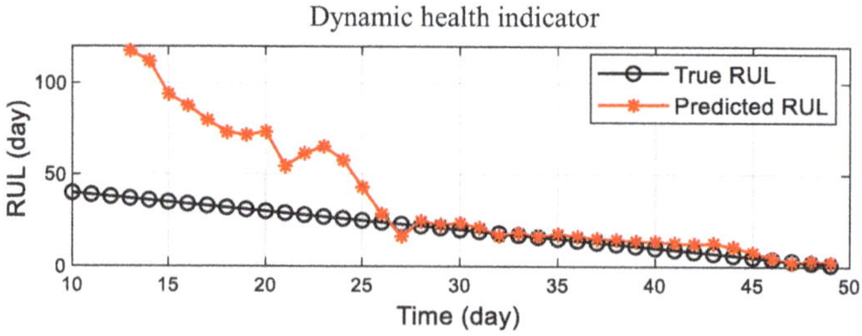

Fig. 6 Predicted RUL based on the dynamic HI

model based on the dynamic HI succeeded in estimating the RULwith a fairly high accuracy from day 27 (based only on 26 degradation data), whereas with the static indicators, the accuracy started to be reasonable only from day 36. This improvement stems from the ability of the proposed dynamic HI selection method to choose the best feature that reflects the degradation trend based on the available data.

4 Conclusion

In this chapter, an interactive data-based prognostic approach is presented to overcome the limitations of historical degradation data of industrial machineries. A DBSCAN classifier is used to detect the degradation of a real shaft bearings system installed in a wind turbine. An original approach is used to estimate the RUL of the system. The RUL prediction is performed using an exponential degradation model. The latter is carried out based on a dynamically selected HI, chosen according to the available degradation data. The proposed approach proved to be efficient compared to the traditional static HI based prognostic approach.

In the following works, the proposed approach will be tested in a more sophisticated way against degradation data of higher complexity.

References

1. Liu J, Wang W, Golnaraghi F (2009) A multi-step predictor with a variable input pattern for system state forecasting, vol 23, pp 1586–1599. https://doi.org/10.1016/j.ymssp.2008.09.006
2. Lei Y (2016) Intelligent fault diagnosis and remaining useful life prediction of rotating machinery. Butterworth-Heinemann
3. Jin X, Que Z, Sun Y, Guo Y, Qiao W (2019) A data-driven approach for bearing, vol 55(4), pp 3394–3401
4. Ben J, Harrath S, Bechhoefer E, Benbouzid M (2017) Online automatic diagnosis of wind turbine bearings progressive degradations under real experimental conditions based on unsupervised machine learning, vol 132, pp 167–181. https://doi.org/10.1016/j.apacoust.2017.11.021
5. Kerroumi S, Chiementin X, Rasolofondraibe L (2013) Dynamic classification method of fault indicators for bearings' monitoring. Mech Ind 14(2):115–120. https://doi.org/10.1051/meca/2013058
6. Farhat MH, Chiementin X, Chaari F, Bolaers F, Haddar M (2020) Digital twin-driven machine learning: ball bearings fault severity classification. Meas Sci Technol
7. Olivares-Mercado J, Aguilar-Torres G, Toscano-Medina K, Sanchez-Perez G, Nakano-Miyatake M, Perez-Me H (2012) Multidimensional features extraction methods in frequency domain. Fourier Transform Signal Process. https://doi.org/10.5772/36704
8. Farhat MH, Hentati T, Chiementin X, Bolaers F, Chaari F, Haddar M (2020) Numerical model of a single stage gearbox under variable regime. Mech Des Struct Mach 1993
9. Gebraeel N, Lawley M, Liu R, Parmeshwaran V (2004) Residual life predictions from vibration-based degradation signals: a neural network approach. IEEE Trans Ind Electron 51(3):694–700. https://doi.org/10.1109/TIE.2004.824875

10. Chen C, Vachtsevanos G, Orchard ME (2010) Machine remaining useful life prediction based on adaptive Neuro-Fuzzy and High-Order particle filtering. In: Annual Conference on Prognostics and Health Management Society (PHM 2010), pp 1–9
11. Wang P, Long Z, Wang G (2020) A hybrid prognostics approach for estimating remaining useful life of wind turbine bearings. Energy Rep 6(1):173–182. https://doi.org/10.1016/j.egyr.2020.11.265
12. Kong X, Yang J (2019) Remaining useful life prediction of rolling bearings based on RMS-MAVE and dynamic exponential regression model. IEEE Access 7:169705–169714. https://doi.org/10.1109/ACCESS.2019.2954915
13. SchSlkopf B, Smola A, Muller K-R (2011) Kernel principal component analysis. In: Artificial Neural Networks (ICANN'97). ICANN 1997. Lecturer Notes in Computer Science vol 3, pp 1–6
14. Baalis Coble J, Baalis J (2010) Trace: Tennessee research and creative exchange merging data sources to predict remaining useful life—an automated method to identify prognostic parameters recommended citation
15. Bechhoefer E, Van Hecke B, He D (2013) Processing for improved spectral analysis. In: PHM 2013—Proceedings of the Annual Conference of Prognostics and Health Management Society, pp 33–38
16. Si XS, Wang W, Chen MY, Hu CH, Zhou DH (2013) A degradation path-dependent approach for remaining useful life estimation with an exact and closed-form solution. Eur J Oper Res 226(1):53–66. https://doi.org/10.1016/j.ejor.2012.10.030

The manufacturer's authorised representative in the EU is Springer Nature Customer Service Centre GmbH, Europaplatz 3, 69115 Heidelberg, Germany. If you have any concerns regarding our products, please contact ProductSafety@springernature.com

Printed and bound by CPI Group (UK) Ltd, Croydon, CR0 4YY
30/09/2026
02225665-0013